Electronic Word of Mouth as a Promotional Technique

Recent years have seen digital advertising grow to the point where it will soon overtake television as the no. 1 advertising medium. In the online environment, consumers interact and share their thoughts on brands and their experiences using them. These electronic word-of-mouth (eWOM) communications have become very important to the success of products. In today's cluttered environment, it is especially important to study how the practice of eWOM advertising operates, and how marketers can influence eWOM in social media and other online sites.

This volume starts with a chapter on the current state of knowledge on eWOM and then turns its attention to current research articles on a variety of eWOM formats. These include the posting of selfies on social media, the influence of review types on consumer perception and purchase intention, the effects of preannouncement messages, and how user-generated content can be used to induce effectiveness of eWOM on social media. The relationship of eWOM to brand building is emphasized in several of the chapters.

This book was originally published as a special issue of the *International Journal of Advertising*.

Shu-Chuan Chu is an Associate Professor and Program Chair of Public Relations and Advertising in the College of Communication at DePaul University, Chicago, USA. Her research interests include social media, electronic word-of-mouth, and international advertising. Her work has been published in leading advertising and marketing journals.

Juran Kim is a Professor of Business Administration at Jeonju University, South Korea. She currently serves as the Associate Executive Secretary for the Global Alliance of Marketing and Management Scholars. Her work has appeared in the *Journal of Advertising*, *International Journal of Advertising*, *Journal of Business Research*, *Journal of Consumer Mediate Communications*, and several other leading publications.

Charles R. Taylor is the John A. Murphy Professor of Marketing at Villanova University, USA. He currently serves as the Editor-in-Chief of the *International Journal of Advertising*. He has published more than 100 peer-reviewed articles and has received the Ivan L. Preston and Flemming Hansen Awards for lifetime contribution to advertising research. He is a past President of the American Academy of Advertising.

Electronic Word of Mouth as a Promotional Technique

New Insights from Social Media

Edited by
**Shu-Chuan Chu, Juran Kim
and Charles R. Taylor**

LONDON AND NEW YORK

First published 2019
by Routledge
2 Park Square, Milton Park, Abingdon, Oxon, OX14 4RN, UK

and by Routledge
52 Vanderbilt Avenue, New York, NY 10017, USA

First issued in paperback 2020

Routledge is an imprint of the Taylor & Francis Group, an informa business

British Library Cataloguing-in-Publication Data
A catalogue record for this book is available from the British Library

ISBN 13: 978-0-367-58476-4 (pbk)
ISBN 13: 978-1-138-36090-7 (hbk)

Typeset in Myriad Pro
by codeMantra

Publisher's Note
The publisher accepts responsibility for any inconsistencies that may have arisen during the conversion of this book from journal articles to book chapters, namely the possible inclusion of journal terminology.

Disclaimer
Every effort has been made to contact copyright holders for their permission to reprint material in this book. The publishers would be grateful to hear from any copyright holder who is not here acknowledged and will undertake to rectify any errors or omissions in future editions of this book.

Contents

CONTENTS

Citation Information

The chapters in this book were originally published in the *International Journal of Advertising*, volume 37, issue 1 (January 2018). When citing this material, please use the original page numbering for each article, as follows:

Chapter 1
The current state of knowledge on electronic word-of-mouth in advertising research
Shu-Chuan Chu and Juran Kim
International Journal of Advertising, volume 37, issue 1 (January 2018) pp. 1–13

Chapter 2
#Me and brands: understanding brand-selfie posters on social media
Yongjun Sung, Eunice Kim and Sejung Marina Choi
International Journal of Advertising, volume 37, issue 1 (January 2018) pp. 14–28

Chapter 3
Understanding the effects of different review features on purchase probability
Su Jung Kim, Ewa Maslowska and Edward C. Malthouse
International Journal of Advertising, volume 37, issue 1 (January 2018) pp. 29–53

Chapter 4
Preannouncement messages: impetus for electronic word-of-mouth
Hao Zhang and Yung Kyun Choi
International Journal of Advertising, volume 37, issue 1 (January 2018) pp. 54–70

Chapter 5
The power of e-WOM using the hashtag: focusing on SNS advertising of SPA brands
Jiye Shin, Heeju Chae and Eunju Ko
International Journal of Advertising, volume 37, issue 1 (January 2018) pp. 71–85

Chapter 6
Do we always adopt Facebook friends' eWOM postings? The role of social identity and threat
Yaeri Kim, Yookyung Park, Youseok Lee and Kiwan Park
International Journal of Advertising, volume 37, issue 1 (January 2018) pp. 86–104

Chapter 7

When brand-related UGC induces effectiveness on social media: the role of content sponsorship and content type
Mikyoung Kim and Doori Song
International Journal of Advertising, volume 37, issue 1 (January 2018) pp. 105–124

Chapter 8

What does the brand say? Effects of brand feedback to negative eWOM on brand trust and purchase intentions
Manu Bhandari and Shelly Rodgers
International Journal of Advertising, volume 37, issue 1 (January 2018) pp. 125–141

Chapter 9

The interconnected role of strength of brand and interpersonal relationships and user comment valence on brand video sharing behaviour
Jameson L. Hayes, Yan Shan and Karen Whitehill King
International Journal of Advertising, volume 37, issue 1 (January 2018) pp. 142–164

For any permission-related enquiries please visit:
http://www.tandfonline.com/page/help/permissions

Notes on Contributors

Manu Bhandari is an Assistant Professor of Strategic Communication in the Department of Communication at Arkansas State University's College of Liberal Arts and Communication, Jonesboro, USA. His research areas include online information processing and electronic word-of-mouth. At Arkansas State, he currently teaches courses in advertising principles, interactive advertising, and media writing.

Heeju Chae (PhD, Yonsei University, Seoul, South Korea) is an Assistant Professor in the Department of International Trade and Commerce at Kyungsung University, Busan, South Korea. Her research interests include social media marketing, social responsibility, sustainable brand marketing, retailing, consumer behavior, and customer equity.

Sejung Marina Choi is a Professor of Advertising in the School of Media and Communication at Korea University, Seoul, South Korea. Her research interests include advertising in new media, branded entertainment, and consumer–brand relationships. In addition to her academic work, she has several years of professional experience working at an advertising agency.

Yung Kyun Choi (PhD, Michigan State University, East Lansing, USA) is a Professor in the Department of Advertising and PR at Dongguk University, Seoul, South Korea. His research interests include message congruency effects, consumer behavior in virtual social environments, and cross-cultural advertising effects. He has received Best Reviewer Award from *International Journal of Advertising*, and received Highly Commended Award from *Internet Research* more recently. His work has appeared in leading advertising, and marketing journals.

Shu-Chuan Chu is an Associate Professor and Program Chair of Public Relations and Advertising in the College of Communication at DePaul University, Chicago, USA. Her research interests include social media, electronic word-of-mouth, and international advertising. Her work has been published in leading advertising and marketing journals including *International Journal of Advertising*, *Journal of Interactive Marketing*, *Journal of Global Marketing*, *Journal of Interactive Advertising*, *Journal of Marketing Communications*, among others.

Jameson L. Hayes is an Assistant Professor of Advertising in the Department of Advertising + Public Relations in the College of Communication & Information Sciences at the University of Alabama, Tuscaloosa, USA. His research interests include emerging media marketing, consumer–brand relationships, and advertising field development.

Eunice Kim, PhD, is an Assistant Professor in the Department of Psychology at Ewha Woman's University, Seoul, South Korea. Her research focuses on consumer and advertising psychology, marketing and persuasive communications in digital media environments, consumer–brand relationships, and media psychology.

Juran Kim is a Professor of Marketing, Business Administration at Jeonju University, South Korea. Her research interest includes digital advertising and sustainable marketing. Her work has appeared in the *Journal of Advertising, International Journal of Advertising, Journal of Business Research, Journal of Consumer Mediate Communications*, and several other leading publications.

Mikyoung Kim (PhD, Michigan State University, East Lansing, USA) is an Assistant Professor in the School of Advertising and Public Relations at Hongik University, Sejong, South Korea. Her research focuses on consumer responses to marketing strategies on various digital media.

Su Jung Kim is an Assistant Professor in the Annenberg School for Communication and Journalism at the University of Southern California. She received her Ph.D. degree in Media, Technology, and Society (MTS) from the School of Communication, and was a Post-doctoral Research Associate in the Spiegel Center for Digital and Database Marketing at Northwestern University. Before joining USC, she was an Assistant Professor in the Greenlee School of Journalism and Communication at Iowa State University. Her research centers on the use of big data in communication, advertising, and marketing communications. Her research interests include cross-platform media use and its effects, social media and electronic word-of-mouth (e-WOM), and mobile customer engagement. Her current projects examine how online product reviews influence the perceptions and behaviors of consumers.

Yaeri Kim (MS, Seoul National University, South Korea) is a PhD Candidate in the Marketing Department at Seoul National University, South Korea. Her research interests include identity-related issues in a consumer behavior context (e.g. brand identity, identity threat, and moral identity), and she recently extended her interest to neuro-marketing.

Karen Whitehill King is the Jim Kennedy Professor of New Media, Josiah Meigs Distinguished Teaching Professor and Professor of Advertising in the Grady College of Journalism and Mass Communication at the University of Georgia, Athens, USA. Her research interests include advertising industry issues, digital media, and health communication.

Eunju Ko (PhD, Virginia Tech, Blacksburg, USA) is a Professor of Fashion Marketing and an Adjunct Professor of Culture & Design Management Program at Yonsei University, Seoul, South Korea. Her research interests include social media marketing, luxury branding, and sustainable & fashion culture management. Her work has appeared in such journals as the *Journal of Business Research, International Marketing Review, Psychology & Marketing*, and the *Journal of Global Fashion Marketing*, among others.

Youseok Lee is an Assistant Professor in the College of Business Administration at Myongji University, Seoul, South Korea. His research interests include social influence (e.g. WOM behavior, herd behavior), marketing strategy, and entertainment marketing.

Edward C. Malthouse is the Theodore R and Annie Laurie Sills Professor of Integrated Marketing Communication, and Industrial Engineering and Management Science, at

Northwestern University, Evanston, USA. He is the Research Director for the Medill IMC Spiegel Digital & Database Research Center. His research interests center on customer engagement and experiences, digital social and mobile media, big data, customer lifetime value models, predictive analytics, unsupervised learning, and integrated marketing communication.

Ewa Maslowska is an Assistant Professor in the Amsterdam School of Communication Research at the University of Amsterdam, Netherlands, where she also earned her PhD degree from persuasive communication program. She completed a postdoc in the Medill IMC Spiegel Digital & Database Research Center at Northwestern University, Evanston, USA. Her research interests center around consumer behavior, advertising, and digital consumer environments. She conducts experimental and data-driven research into the dynamics of consumer engagement.

Kiwan Park is a Professor of Marketing at Seoul National University Business School, South Korea. With expertise in psychology, his current research addresses important theoretical topics in consumer behavior, such as prosocial behavior, social cognition, branding, and persuasion. He has published more than 30 articles in top academic journals, including *Journal of Consumer Research, Journal of Consumer Psychology, and Organizational Behavior and Human Decision Processes* among others. He holds a Ph.D. in Business Administration (Marketing) from the University of Michigan (Ann Arbor), and both a B.B.A. and M.S. from Seoul National University.

Yookyung Park (MS, Seoul National University, South Korea) is a PhD Candidate in the Marketing Department at Seoul National University, South Korea. Her research interests include consumer psychology, promotion management, and online consumer behavior.

Shelly Rodgers is a Professor of Strategic Communication in the School of Journalism at the University of Missouri, Columbia, USA. Rodgers is nationally ranked as a leading Internet advertising researcher and is among the 10 most cited Internet advertising, marketing, and communications scholars in the US. Her research areas include Internet advertising, new technology, and health communication.

Yan Shan is an Assistant Professor in the Journalism Department in the College of Liberal Arts at California Polytechnic State University, San Luis Obispo, USA. Her research interests include media strategy and advertising effects and the impact of emerging communications technologies on advertising and public relations.

Jiye Shin (MS, Yonsei University, Seoul, South Korea) is a graduate from the Department of Clothing & Textiles at Yonsei University, Seoul, South Korea. Her research interests include online shopping, social media marketing, and brand image.

Doori Song (PhD, University of Florida, Gainesville, USA) is an Associate Professor of Marketing in the Warren P. Williamson, Jr. College of Business Administration at Youngstown State University, USA. His research focuses on consumer psychology and behavior.

Yongjun Sung is a Professor of Psychology in the Department of Psychology at Korea University. Dr. Sung's research focuses on self-expression, branding, Internet of Things (IOT), AI, and a variety of consumer psychology issues. He has authored or co-authored over 80 articles in leading refereed journals including *International Journal of Advertising, Journal of Advertising, Journal of Consumer Psychology, Journal of Cross-Cultural*

Psychology, Psychology & Marketing, Journal of Business Research, Marketing Letters, and *Journal of Public Policy and Marketing,* among others.

Charles R. Taylor is the John A. Murphy Professor of Marketing at Villanova University, USA. He currently serves as the Editor-in-Chief of the International Journal of Advertising. He has published more than 100 peer-reviewed articles and has received the Ivan L. Preston and Flemming Hansen Awards for lifetime contribution to advertising research. He is a past President of the American Academy of Advertising.

Hao Zhang is an associate professor and the chair professor in the Department of Marketing at Northeastern University, Shenyang, China. His research interests include new product development, digital marketing, and integrated marketing communication. His current research project examines how customer value co-creation behaviors influence new product development.

The current state of knowledge on electronic word-of-mouth in advertising research

Shu-Chuan Chu and Juran Kim

ABSTRACT

With the development of new and digital media, consumers are increasingly giving, seeking, and sharing their brand-related experiences via online channels that lead to electronic word-of-mouth (eWOM) communication. Undoubtedly, eWOM has a powerful impact on advertising decisions, and the practice of eWOM advertising has received much attention from advertisers, policy-makers, and the academic community. It is clear as well that, eWOM is believed to influence and shape the future of advertising. Therefore, the purpose of this article is threefold. First, this article provides an overview on the current state of eWOM research, with a focus on how this research has evolved in the advertising literature. Second, we identify and characterize four important trends in the eWOM advertising literature: eWOM and viral advertising, effects of eWOM, drivers of eWOM, and new technologies and eWOM platforms. Third, we develop a research agenda by providing directions for future research as well as implications for advertisers and policy-makers.

Introduction

The advent of new media technologies has enabled consumers to use online and digital communications as a channel for information exchange. As digital media continue to evolve, research investigating how the Internet and new technologies affect electronic word-of-mouth (eWOM) behaviour has gained unprecedented levels of attention over the past few decades (Brown, Broderick, and Lee 2007; Dellarocas 2003; Dwyer 2007; Goldsmith and Horowitz 2006; Keller 2007). Recently, interest has also grown in the nature and effect of eWOM and its relationship with advertising (Lee, Ham, and Kim 2013; Liu-Thompkins 2012; Phelps et al. 2004; Precourt 2015; San José-Cabezudo and Camarero-Izquierdo 2012; Smith et al. 2007; Strutton, Taylor, and Thompson 2011; Taylor 2017a).

eWOM refers to 'any positive or negative statement made by potential, actual, or former customers about a product or company, which is made available to a multitude of people and institutions via the Internet' (Hennig-Thurau et al. 2004, 39). Simply defined, eWOM involves the behaviour of exchanging marketing information among consumers in online environments or via new technologies (e.g. mobile communication). Previous research has demonstrated the many different channels where eWOM occurs, such as discussion boards (e.g. Fong and Burton 2006; Steyer, Garcia-Bardidia, and Quester 2006), product review sites (e.g. Chevalier and Mayzlin 2006; Maslowska, Malthouse, and Bernritter 2017; Wang, Cunningham, and Eastin 2015), virtual consumer communities (e.g. Hung and Li 2007), emails (e.g. Chiu et al. 2007; Phelps et al. 2004), blogs (e.g. Thorson and Rodgers 2006), and social media sites like Facebook, Twitter, Snapchat, and WeChat (Chu and Kim 2011; Colliander, Dahl'en, and Modig 2015; Jin and Phua 2014).

The interactive nature of digital media allows consumers to conveniently give and seek information about product experiences, thereby affecting consumers' purchase intentions and consumption decisions (Prendergast, Ko, and Yuen 2010). Because eWOM is generated and disseminated among peer consumers without commercial interests, it has been recognized as a more trustworthy source of information compared to advertiser-generated messages and thus has a greater influence on consumers' product evaluation (Brown, Broderick, and Lee 2007). Moreover, the skyrocketing rise of social media in recent years gives consumers unlimited access to a great amount of information regarding product and brand choices, facilitating eWOM behaviours (Araujo, Neijens, and Vliegenthart 2017; Chu and Sung 2015; Jin and Phua 2014; Yoon, Polpanumas, and Park 2017). Social media provide consumers with another channel to search for reliable, unbiased product information and at the same time allows advertisers to interact with their consumers by participating in eWOM. As recent marketing disasters from Pepsi and United Airlines highlight the importance of eWOM research (see Taylor 2017a), it is time to think about what we know about eWOM and what areas of eWOM need more work.

Accordingly, this article aims to review the current state of knowledge regarding eWOM as pertains to advertising and provides directions for future research. To achieve this goal, research of eWOM literature was conducted with major advertising journals, including *Journal of Advertising, Journal of Advertising Research, International Journal of Advertising, Journal of Interactive Advertising*, and *Journal of Current Issues & Research in Advertising*. Relevant literature published in marketing and communication journals are also included to expand the scope of the review. The purpose of this article is not to provide a comprehensive literature review on eWOM in advertising research. What we attempt is to identify trends over the past decades and make recommendations for future research agendas. eWOM is not a new subject, but a close examination of the state of eWOM in advertising research provides valuable insights for academic researchers and educators as well as important implications for advertisers and marketers wanting to leverage the power of eWOM in their advertising campaigns.

Brief history of the electronic word-of-mouth literature

eWOM is an expanded form of word-of-mouth (WOM), which refers to information consumers obtain from interpersonal sources such as family and friends (Engel, Blackwell, and Kegerreis 1969; Gilly et al. 1998). The influence of WOM on consumer decision-making has been well documented in marketing and consumer behaviour literature since the 1960s (Goldsmith and Clark 2008; Feick and Price 1987). For example, WOM has been found to influence the speed and rate of innovation diffusion (Mahajan, Muller, and Srivastava 1990) and is key for product adoption by impacting the flow of information in social networks (Frenzen and Nakamoto 1993). Because consumers generally perceive interpersonal sources to be more credible than commercial messages (Feick and Price 1987), WOM has been recognized as an important source of information in changing consumer attitude and behaviour related to products and services (Engel, Blackwell, and Kegerreis 1969; Grewal, Cline, and Davies 2003; Moon, Costello, and Koo 2017).

As the Internet had a revolutionary impact on consumers' daily life in the mid-1990s, WOM evolved into online space, and the concept of eWOM was introduced. Since the early 2000s, academic research on eWOM has steadily emerged in disciplines such as management (Dellarocas 2003), marketing (Dwyer 2007; Godes and Mayzlin 2004), electronic commerce (Balasubramanian and Mahajan 2001; Goldsmith and Horowitz 2006), and advertising (Fong and Burton 2006). In today's world of fragmented advertising media, eWOM is recognized as an important advertising technique that has the potential to drive product success. Many advertisers are interested in applying the power of eWOM advertising (Phelps et al. 2004) to promote products and services as well as to develop brand loyalty. In particular, viral marketing or eWOM advertising demonstrates marketers' attempts to harness eWOM as an advertising tool (Phelps et al. 2004; Porter and Golan 2006). As a result, eWOM has been recognized as an important area of advertising research that attracts attention from both scholars and practitioners.

With the rapid development of social media since the early 2000s and its popularity through the years, a growing body of research attempts to examine how eWOM works in social media and how it is different from eWOM communicated via other channels (Chu and Kim 2011; Eisingerich et al. 2015; Jansen et al. 2009; Jin and Phua 2014; Yoon, Polpanumas, and Park 2017). Today, many companies invest substantial efforts on social media marketing by sparking positive eWOM communications and accelerating its dissemination (Chu, Chen, and Sung 2016; Fogel 2010). Because the personal networks or circles (e.g. WeChat's friend circles) are pre-existing and available on social media sites such as Facebook, Twitter, and Instagram, eWOM communicated via social media has become an influential source of product information for consumers. eWOM within social media is perceived as more trustworthy than it is when generated from other channels (e.g. product review sites) and when the sources of information are members in consumers' personal networks. As social media become an effective vehicle for eWOM advertising, systematic research examining the impact of eWOM in social media on consumer decision-making has emerged as a popular topic in the current advertising literature (Araujo, Neijens, and Vliegenthart 2017; Colliander, Dahl'en, and Modig 2015; Knoll 2016). In order to provide an overview of the state of knowledge on eWOM in advertising research, the next section presents our observations on trends in eWOM research over the past few decades.

Trends in electronic word-of-mouth research

eWOM and viral advertising

Since the early 2000s, eWOM has been frequently used to examine the topic of viral advertising or viral marketing (Eckler and Bolls 2011; Liu-Thompkins 2012; Phelps et al. 2004; Porter and Golan 2006). Phelps et al.'s (2004) seminal work is one of the first studies examining consumer responses and motivations to pass-along email messages concerning products or services. Using a three-phase research method, Phelps et al. (2004) considered viral marketing as a form of eWOM advertising and identified entertainment and social motivations as top-rated reasons for pass-along behaviours. Phelps et al.'s (2004) study provides a starting point for eWOM research in relation to advertising and encourages future examination on pass-along email behaviours. Porter and Golan's (2006) later study further defined viral advertising as a division of eWOM and represented the first empirical effort to distinguish between viral and television advertising. Specifically, Porter and Golan (2006) defined viral advertising as 'unpaid peer-to-peer communication of provocative content originating from an identified sponsor using the Internet to persuade or influence an audience to pass-along the content to others (33).' Both Phelps et al.'s (2004) and Porter and Golan's (2006) studies open opportunities and challenges to research eWOM and its relationship with advertising.

More recently, Cho, Huh, and Faber (2014) defined viral advertising as 'marketer-initiated eWOM strategies that use specially crafted messages designed to be passed along or spread by consumers' (100). This definition is in line with Porter and Golan's (2006) conceptualization of viral advertising that illustrates the important aspects of peer-to-peer communication and eWOM behaviour. Cho, Huh, and Faber's (2014) study found that a viral ad sent by a trusted sender can overcome the disadvantage of a less trusted advertiser, suggesting that sender trust plays an important role in the multistage process of viral advertising effects. Other eWOM research in relation to viral advertising includes, for example, Shan and King (2015) and Hayes and King (2014). Shan and King (2015) found that a strong consumer–brand relationship increases the effectiveness of eWOM (e.g. passing along viral advertising messages) on social networking sites. Similarly, results from Hayes and King's (2014) study provide evidence that brand relationships intertwine with sharing motivations to impact consumers' decisions to pass-along viral video ads within social networking sites.

Overall, research in this area suggests that viral advertising can capitalize on the impact of eWOM (Hayes and King 2014) and considers social and individual factors that lead to pass-along behaviours. The advent of mobile technologies and social media applications has made it easier for consumers to forward viral ads without time and location constraint. Because viral advertising is considered as a form of eWOM, it is important to understand what motivates consumers to pass-along viral ads and share content about brands in a variety of digital channels. Meanwhile, theoretical studies examining the outcomes of viral ads are needed in order to advance our knowledge regarding the role of viral advertising in the eWOM process.

Effects of eWOM

The great amount of research on eWOM focuses on how it works (Craig, Greene, and Versaci 2015; Fulgoni and Lipsman 2015) and its effects on important marketing outcomes such as purchase decisions (De Bruyn and Lilien 2008; López and Sicilia 2014; Prendergast, Ko, and Yuen 2010; See-To and Ho 2014) and sales (Chevalier and Mayzlin 2006; Godes and Mayzlin 2004; Huang et al. 2014). For example, Senecal and Nantel (2004) examined whether online product recommendations influence consumers' product evaluation and demonstrated that online product recommendations, a popular form of eWOM, have an influential impact on consumers' product-related decisions. Their study found that individuals who were exposed to product recommendations were more likely to select the recommended products than those who did not receive any recommendations. Chevalier and Mayzlin's (2006) study showed that positive product reviews lead to increased sales of that product, indicating the powerful role that eWOM plays in the decision-making process in an e-commerce environment.

The research stream on the effects of eWOM also involves message characteristics and source credibility (Jin and Phua 2014; Kareklas, Muehling, and Weber 2015; Lee, Rodgers, and Kim 2009; Wang, Cunningham, and Eastin 2015). For example, Lee, Rodgers, and Kim (2009) examined the effects of valence and extremity of eWOM via consumer product reviews on attitude toward the brand and website. Their research supported the positive relationship between extremely positive reviews and attitude toward the brand. Furthermore, results from their experiments confirmed the negativity and extremity effect, suggesting that extremely negative reviews produced a strongest influence on attitude toward the brand compared to either extremely positive reviews or moderately negative product reviews. Wang, Cunningham, and Eastin (2015) study also extended the research on online consumer reviews by exploring the influence of eWOM message content characteristics. In particular, the message valence of online consumer reviews (positive, negative, and neutral) was found to have an influence on consumers' attitudes toward the review and product, perceived source credibility, and purchase intention (Wang, Cunningham, and Eastin 2015).

As eWOM research continues to grow in popularity, research examining the effects and drivers of eWOM has emerged at the same time (e.g. Kim, Cheong, and Kim 2017). Understanding the effects and drivers of eWOM could provide useful implications for advertising scholars and practitioners. From a theoretical perspective, identifying factors leading to eWOM behaviours and ultimate eWOM outcomes could provide a complete picture of the diffusion process in an online environment. As for practical implications, marketers could use research findings to develop effective eWOM campaigns that reach their marketing goals. The next section discusses recent trends of eWOM research as pertains to its drivers and motives.

Drivers of eWOM

Research examining antecedents and motives of eWOM behaviour is a subject that received rising attention from researchers in recent years (Chu and Sung 2015; Shi and Wojnicki 2014). In the marketing and communication literature, a few studies have examined factors that drive consumers' eWOM behaviour (Balasubramanian and Mahajan 2001; Hennig-Thurau et al. 2004). In their influential study, Hennig-Thurau et al. (2004) examined the underlying process that impacts consumers' use of consumer-opinion platforms and identified factors that lead to eWOM behaviour. Specifically, consumers' desire for social interaction, desire for economic incentives, concern for other consumers, and the potential approval utility are important motives that drive eWOM behaviour via consumer-opinion platforms. Hennig-Thurau et al.'s (2004) findings echo prior eWOM literature on virtual community (Balasubramanian and Mahajan 2001). Recently, Lien and Cao (2014) examined the relationships among WeChat users' motivations, trust, attitudes, and eWOM and found that entertainment, sociality, information, and trust positively lead to

WeChat users' attitudes, and their trust and attitudes significantly influence positive eWOM (Lien and Cao 2014).

In the advertising literature on eWOM, a growing number of studies have contributed to our understanding of eWOM advertising by exploring the motives of eWOM behaviour (e.g. Chatterjee 2011; Chu and Kim 2011; Wallace et al. 2014). Applying social capital theory and the information-processing perspective, San José-Cabezudo and Camarero-Izquierdo (2012) investigated how structural, relational, and cognitive social capital; the message characteristics (source, subject, and content); motivational factors (entertainment, sharing, and benefit); and the situational context impact the intention of opening and forwarding email messages. Their study confirmed that individuals' social capital, as well as message source and content, are determinants that affect opening and forwarding emails. Furthermore, entertainment and share motivations are relevant determinants of forwarding while the situational characteristics of the context influence intention to open messages (San José-Cabezudo and Camarero-Izquierdo 2012). Similarly, Chatterjee (2011) investigated the drivers of an influencer's decision on new product recommending and the recipient's decision on referral behaviour on social networking sites. Research has demonstrated the important role of an influencer's activity on a social networking site, brand message source, and recipient type in the eWOM process (Chatterjee 2011). In summary, continued attention should be paid on the drivers of eWOM, as motives to participate in eWOM are likely to vary based on the components of eWOM and platforms.

New technologies and eWOM platforms

Another trend that is identified in eWOM in advertising research involves new technologies and eWOM platforms/channels (Levy and Gvili 2015). For example, Lee and Youn (2009) explored how eWOM platforms influence consumer product judgment and examined how the valence of eWOM moderated the relationship between platforms and product judgment. Interestingly, Lee and Youn's (2009) study suggests that different online platforms to which eWOM is posted produced distinct effects on consumers' judgment about products, with reviews on an independent review website or a brand's website being more likely to lead to product recommendations among consumers than reviews posted on the personal blog. The moderating role of the valence of eWOM was also confirmed in their study. Okazaki (2009) investigated consumer participation in eWOM by comparing eWOM platforms between PC and mobile Internet. Drawing from the social influence model, Okazaki (2009) found that social identity, motivations, inherent novelty seeking, and opinion leadership are factors that lead to eWOM behaviours. His study also illustrates that consumers who engaged in mobile-based eWOM showed higher levels of perceptions on social intention, intrinsic enjoyment, and cognitive social identity than those who participated in eWOM via a PC platform.

As mentioned earlier, social media have provided consumers with an unparalleled venue to participate in eWOM easily and quickly. Social media allow consumers to share their brand-related experiences with contacts in their social network on a global scale, accelerating eWOM communication (Chen, Fay, and Wang 2011; De Veirman, Cauberghe, and Hudders 2017). As social media have become one of the most widely used eWOM formats, research exploring the potential influence of eWOM in social media on advertising strategies and brand communications has gained vast attention in recent years (Carr and

Hayes 2014; Colliander, Dahl'en, and Modig 2015; Eisingerich et al. 2015; Wallace et al. 2014; Yoon, Polpanumas, and Park 2017). For example, Jansen and his colleagues (2009) argue that Twitter is a new form of eWOM for consumers to exchange information, which provides implications for digital advertising strategies. Jin and Phua (2014) focused on celebrities' tweets about brands and examined how eWOM distributed via Twitter impacts consumers' source credibility perception, buying intention, and identification with celebrities. Recent study also explored the impact of Twitter-based eWOM on moviegoers' decisions and film revenues (Yoon, Polpanumas, and Park 2017).

Recognizing the important role of social media in building and maintaining online social relationships, Chu and Kim (2011) examined the determinants of consumers' engagement in eWOM via social networking sites. In their study, Chu and Kim (2011) identified key social relationship factors that are associated with eWOM within online social websites and found that tie strength, trust, and normative and informational influence are positively related to consumers' engagement in eWOM. Yet, homophily was found to negatively influence consumers' willingness to participate in eWOM in online social space. This study provides important insights regarding the influence of social relationship factors on consumers' eWOM behaviour in social networking sites and enhances our theoretical understanding of the underlying process of eWOM. Recently, Farías (2017) expanded the research on eWOM in social networking sites into the context of Chile. In sum, eWOM research related to new technologies and eWOM platforms is still emerging and growing. As new and digital technologies are changing rapidly, there exists a number of opportunities for advertising research that could advance our theoretical knowledge on eWOM and provide relevant information to the advertising industry.

Directions for future research on eWOM

How effectively academics and practitioners deal with the challenges of eWOM will define the success of advertising for some time to come. Three major future research directions can be used to expand upon current eWOM studies: (1) the effects of eWOM using a new ad type, native advertising, (2) various new technology-driven settings, and (3) viral advertising for prosocial relationships.

The current eWOM research has dealt with how eWOM works and its effects on important marketing outcomes. The emergence of native advertising could provide new research opportunities for eWOM effects. Native advertising has recently gained attention, as it assists in overcoming ad clutter and boosting diminishing ad revenue (Taylor 2017b; Wojdynski and Evans 2016; Wojdynski et al. 2017). Native ads refer to paid ads that are interrelated with the content, integrated into the design, and consistent with the platform behaviour that viewers feel they fit into (Interactive Advertising Bureau 2013). Native advertising is a promising new venue for advertisers because it provides more eWOM opportunities than do traditional ads, which users are likely to avoid or ignore. Native advertising could provide future research directions for eWOM because viewers may not be aware that they are viewing paid content (Carlson 2015; Taylor 2017b), including how eWOM works in the types of overt or covert advertising based on influences of source credibility. These native ads also may exhibit positive effects or negative effects on eWOM.

To date, the eWOM literature has focused on new technologies and eWOM platforms. Newer platforms such as virtual reality (VR), augmented reality, and mixed reality offer advanced prospects for eWOM. When consumers learn about products and consume brands, different experiences arise in a variety of technology-driven settings (Daugherty, Li, and Biocca 2008; Li, Daugherty, and Biocca 2001, 2002; Yim, Chu, and Sauer 2017). Among the new technology-driven platforms currently available, the importance of VR is growing. VR refers to a real or simulated environment in which a perceiver experiences a telepresence (Steuer 1992). VR technology allows the user's subjective sense of being there in a virtual setting to arise in the mind (Choi, Yoon, and Taylor 2015; Kober and Neuper 2013). In 2016, VR market revenue was $1.8 billion. This revenue is expected to increase 168% to $5 billion in 2017, while the global VR market is expected to reach $38 billion in value within three years (Statista 2017).

It is recommended that future studies investigate the influence of the VR presence experience on eWOM. More specifically, a study on the sense of being in a VR setting that stems from a stereoscopic presentation, or a realistic and detailed design (Freeman et al. 1999), can be conducted to determine its influence on eWOM. Another recommended future research direction is investigations into the antecedents and consequences of technology-driven eWOM. This is because a VR user's individual experiences and responses to the VR technology can generate different eWOM effects between different people. How the various effects of individual differences boost eWOM from different VR experiences may also be a new research direction.

Previous eWOM research examined the topic of viral marketing. The studies of this topic can be extended to eWOM in prosocial relationships, allowing people to provide help and resources to others. Homophily is a stronger indicator of eWOM success than the relationships between disseminators and recipients. This is because homophily plays a more vital role in prosocial relationships. Personal value, such as egoism versus altruism orientation, may generate a strong effect on the receptiveness to self-framing versus other-framing (Vaish, Liao and Bellotti 2018). In the prosocial context, homophily enables persuasion. The reason for this is that recipients tend to follow the sender's choices, because they share similar values. Homophily also influences the sender's choice of message for the recipient because the sender better recognizes which message framing type is more persuasive for the recipient (Vaish, Liao and Bellotti 2018). Value orientation and the degree of similarity may direct future research on eWOM in the prosocial context.

These three future research directions (i.e. the effects of eWOM using a new ad type [native advertising], a variety of new technology-driven settings, and viral advertising for prosocial relationships) offer a fruitful venue for converting eWOM, as one of the most persistent concerns of the global advertising industry, into an opportunity to ensure that business is both profitable and viable.

Conclusion

New technologies and digital advertising media are continually evolving in recent years. The rapid change in the advertising media landscape has brought revolutionary opportunities for marketers to engage with their consumers. Despite increased attention having been paid to issues of eWOM advertising over the past decades (Cheung and Thadani

2012), research is scant in many important areas. Although the review of literature is not intended to be exhaustive, it identifies important trends in eWOM research and provides recommendations for additional work. As the prevalence of the Internet, social media, and mobile communications continues to rise, the importance of eWOM in advertising and marketing communications will only intensify (Araujo, Neijens, and Vliegenthart 2017; Colliander, Dahl'en, and Modig 2015; Craig, Greene, and Versaci 2015). Advertisers need to consistently monitor eWOM communication with new and novel methods and take advantage of eWOM and viral advertising. Meanwhile, policy-makers should develop effective guidelines that encourage ethical eWOM advertising and promote consumer protection (WOMMA 2017). In conclusion, eWOM has important implications for advertisers, industry leaders, consumers, researchers, and policy-makers, and it presents a window of opportunity for continued academic research.

Disclosure statement

No potential conflict of interest was reported by the authors.

References

Araujo, T., P. Neijens, and R. Vliegenthart. 2017. Getting the word out on Twitter: The role of influentials, information brokers and strong ties in building word-of-mouth for brands. *International Journal of Advertising* 36, no. 3: 496–513.

Balasubramanian, S. and V. Mahajan. 2001. The economic leverage of the virtual community. *International Journal of Electronic Commerce* 5, no. 3: 103–38.

Brown, J., A.J. Broderick, and N. Lee. 2007. Word of mouth communication within online communities: Conceptualizing the online social network. *Journal of Interactive Marketing* 21, no. 3: 2–20.

Carlson, M. 2015. When news sites go native: Redefining the advertising–editorial divide in response to native advertising. *Journalism* 16, no. 7: 849–65.

Carr, C.T. and R.A. Hayes. 2014. The effect of disclosure of third-party influence on an opinion leader's credibility and electronic word of mouth in two-step flow. *Journal of Interactive Advertising* 14, no. 1: 38–50.

Chatterjee, P. 2011. Drivers of new product recommending and referral behaviour on social network sites. *International Journal of Advertising* 30, no. 1: 77–101.

Chen, Y., S. Fay, and Q. Wang. 2011. The role of marketing in social media: How online consumer reviews evolve. *Journal of Interactive Marketing* 25, no. 2: 85–94.

Cheung, C.M.K. and D.R. Thadani. 2012. The impact of electronic word-of-mouth communication: A literature analysis and integrative model. *Decision Support Systems* 54, no. 1: 461–70.

Chevalier, J.A. and D. Mayzlin. 2006. The effect of word of mouth on sales: Online book reviews. *Journal of Marketing Research* 43, no. 3: 345–54.

Chiu, H.C., Y.C. Hsieh, Y.H. Kao, and M. Lee. 2007. The determinants of email receivers' disseminating behaviors on the Internet. *Journal of Advertising Research* 47, no. 4: 524–34.

Cho, S., J. Huh, and R.J. Faber. 2014. The influence of sender trust and advertiser trust on multistage effects of viral advertising. *Journal of Advertising* 43, no. 1: 100–14.

Choi, Y.K., S. Yoon, and C.R. Taylor. 2015. How character presence in advergames affects brand attitude and game performance: A cross-cultural comparison. *Journal of Consumer Behaviour* 14, no. 6: 357–65.

Chu, S.C., H.T. Chen, and Y. Sung. 2016. Following brands on twitter: An extension of theory of planned behavior. *International Journal of Advertising* 35, no. 3: 421–37.

Chu, S.C. and Y. Kim. 2011. Determinants of consumer engagement in electronic word-of-mouth (eWOM) in social networking sites. *International Journal of Advertising* 30, no. 1: 47–75.

Chu, S.C. and Y. Sung. 2015. Using a consumer socialization framework to understand electronic word-of-mouth (eWOM) group membership among brand followers on Twitter. *Electronic Commerce Research and Applications* 14, no. 4: 251–60.

Colliander, J., M. Dahl'en, and E. Modig. 2015. Twitter for two: Investigating the effects of dialogue with customers in social media. *International Journal of Advertising* 34, no. 2: 181–94.

Craig, C.S., W.H. Greene, and A. Versaci. 2015. E-word of mouth: Early predictor of audience engagement: How pre-release "e-WOM" drives box-office outcomes of movies. *Journal of Advertising Research* 55, no. 1: 62–72.

Daugherty, T., H. Li, and F. Biocca. 2008. Consumer learning and the effects of virtual experience relative to indirect and direct product experience. *Psychology & Marketing* 25, no. 7: 568–86.

De Bruyn, A. and G.L. Lilien. 2008. A multi-stage model of word-of-mouth influence through viral marketing. *International Journal of Research in Marketing* 25, no. 3: 151–63.

Dellarocas, C. 2003. The digitization of word-of-mouth: Promise and challenge of online feedback mechanisms. *Management Science* 49, no. 10: 1407–24.

De Veirman, M., V. Cauberghe, and L. Hudders. 2017. Marketing through Instagram influencers: The impact of number of followers and product divergence on brand attitude. *International Journal of Advertising* 36, no. 5: 798–828.

Dwyer, P. 2007. Measuring the value of electronic word of mouth and its impact in consumer communities. *Journal of Interactive Marketing* 21, no. 2: 63–79.

Eckler, P. and P. Bolls. 2011. Spreading the virus: Emotional tone of viral advertising and its effect on forwarding intentions and attitudes. *Journal of Interactive Advertising* 11, no. 2: 1–11.

Eisingerich, A.B., H.E. H. Chun, Y. Liu, H. M. Jia, and S.J. Bell. 2015. Why recommend a brand face-to-face but not on Facebook? How word-of-mouth on online social sites differs from traditional word-of-mouth. *Journal of Consumer Psychology* 25, no. 1: 120–8.

Engel, J.F., R.D. Blackwell, and R.J. Kegerreis. 1969. How information is used to adopt an innovation. *Journal of Advertising Research* 9, no. 4: 3–8.

Farías, P. 2017. Identifying the factors that influence eWOM in SNSs: The case of Chile. *International Journal of Advertising* 36, no. 6: 852–69.

Feick, L.F. and L.L. Price. 1987. The market maven: A diffuser of marketplace information. *Journal of Marketing* 51, no. 1: 83–97.

Fogel, S. 2010. Issues in measurement of word of mouth in social media marketing. *International Journal of Integrated Marketing Communications* 2: 54–60.

Fong, J. and S. Burton. 2006. Electronic word-of-mouth: A comparison of stated and revealed behavior on electronic discussion boards. *Journal of Interactive Advertising* 6, no. 2: 61–70.

Freeman, J., S.E. Avons, D.E. Pearson, and W.A. IJsselsteijn. 1999. Effects of sensory information and prior experience on direct subjective ratings of presence. *Presence: Teleoperators and Virtual Environments* 8, no.1: 1–13.

Frenzen, J. and K. Nakamoto. 1993. Structure, cooperation, and the flow of market information. *Journal of Consumer Research* 20, no. 3: 360–75.

Fulgoni, G.M. and A. Lipsman. 2015. Digital word of mouth and its offline amplification: A holistic approach to leveraging and amplifying all forms of WOM. *Journal of Advertising Research* 55, no. 1: 18–21.

Gilly, M.C., J.L. Graham, M.F. Wolfinbarger, and L.J. Yale. 1998. A dyadic study of interpersonal information search. *Journal of the Academy of Marketing Science* 26, no. 2: 83–100.

Godes, D. and D. Mayzlin. 2004. Using online conversations to study word of mouth communication. *Marketing Science* 23, no. 4: 545–60.

Goldsmith, R.E. and R.A. Clark. 2008. An analysis of factors affecting fashion opinion leadership and fashion opinion seeking. *Journal of Fashion Marketing & Management* 12, no. 3: 308–22.

Goldsmith, R.E. and D. Horowitz. 2006. Measuring motivations for online opinion seeking. *Journal of Interactive Advertising* 6, no. 2: 1–16.

Grewal, R., T.W. Cline, and A. Davies. 2003. Early-entrant advantage, word-of-mouth communication, brand similarity, and the consumer decision-making process. *Journal of Consumer Psychology* 13, no. 3: 187–97.

Hayes, J.L. and K.W. King. 2014. The social exchange of viral ads: Referral and coreferral of ads among college students. *Journal of Interactive Advertising* 14, no. 2: 98–109.

Hennig-Thurau, T., K.P. Gwinner, G. Walsh, and D.D. Gremler. 2004. Electronic word-of-mouth via consumer-opinion platforms: What motivates consumers to articulate themselves on the Internet? *Journal of Interactive Marketing* 18, no. 1: 38–52.

Huang, L., J. Zhang, H. Liu, and L. Liang. 2014. The effect of online and offline word-of-mouth on new product diffusion. *Journal of Strategic Marketing* 22, no. 2: 177–89.

Hung, K.H. and S.Y. Li. 2007. The influence of eWOM on virtual consumer communities: Social capital, consumer learning, and behavioral outcomes. *Journal of Advertising Research* 47, no. 4: 485–95.

Interactive Advertising Bureau. 2013. *The native advertising playbook.* (December 4), http://www.iab.net/media/file/IAB-Native-Advertising-Playbook2.pdf (accessed September 6, 2017).

Jansen, B.J., M. Zhang, K. Sobel, and A. Chowdury. 2009. Twitter power: Tweets as electronic word of mouth. *Journal of the American Society for Information Science and Technology* 60, no. 11: 2169–88.

Jin, S.A.A. and J. Phua. 2014. Following celebrities' tweets about brands: The impact of Twitter-based electronic word-of-mouth on consumers' source credibility perception, buying intention, and social identification with celebrities. *Journal of Advertising* 43, no. 2: 181–95.

Kareklas, I., D.D. Muehling, and T.J. Weber. 2015. Reexamining health messages in the digital age: A fresh look at source credibility effects. *Journal of Advertising* 44, no. 2: 88–104.

Keller, E. 2007. Unleashing the power of word-of-mouth: Creating brand advocacy to drive growth. *Journal of Advertising Research* 47 (December): 448–52.

Kim, K., Y. Cheong, and H. Kim. 2017. User-generated product reviews on the Internet: The drivers and outcomes of the perceived usefulness of product reviews. *International Journal of Advertising* 36, no. 2: 227–45.

Knoll, J. 2016. Advertising in social media: A review of empirical evidence. *International Journal of Advertising* 35, no. 2: 266–300.

Kober, S.E. and C. Neuper. 2013. Personality and presence in virtual reality: Does their relationship depend on the used presence measure? *International Journal of Human-Computer Interaction* 29, no.1: 13–25.

Lee, J., C.D. Ham, and M. Kim. 2013. Why people pass along online video advertising: From the perspectives of the interpersonal communication motives scale and the theory of reasoned action. *Journal of Interactive Advertising* 13, no. 1: 1–13.

Lee, M., S. Rodgers, and M. Kim. 2009. Effects of valence and extremity of eWOM on attitude toward the brand and Website. *Journal of Current Issues & Research in Advertising* 31, no. 2: 1–11.

Lee, M. and S. Youn. 2009. Electronic word of mouth (eWOM): How eWOM platforms influence consumer product judgement. *International Journal of Advertising* 28, no. 3: 473–99.

Levy, S. and Y. Gvili. 2015. How credible is e-word of mouth across digital-marketing channels? The roles of social capital, information richness, and interactivity. *Journal of Advertising Research* 55 (March): 95–109.

Li, H., T. Daugherty, and F. Biocca. 2001. Characteristics of virtual experience in electronic commerce: A protocol analysis. *Journal of Interactive Marketing* 15, no. 3: 13–30.

Li, H., T. Daugherty, and F. Biocca. 2002. Impact of 3-D advertising on product knowledge, brand attitude, and purchase intention: The mediating role of presence. *Journal of Advertising* 31, no. 3: 43–57.

Lien, C.H. and Y. Cao. 2014. Examining WeChat users' motivations, trust, attitudes, and positive word-of-mouth: Evidence from China. *Computers in Human Behavior* 41: 104–11.

Liu-Thompkins, Y. 2012. Seeding viral content: The role of message and network factors. *Journal of Advertising Research* 52 (December): 465–78.

López, M. and M. Sicilia. 2014. eWOM as source of influence: The impact of participation in eWOM and perceived source trustworthiness on decision making. *Journal of Interactive Advertising* 14, no. 2: 86–97.

Mahajan, V., E. Muller, and R. Srivastava. 1990. Determination of adopter categories using innovation diffusion models. *Journal of Marketing Research* 27, no. 2: 37–50.

Maslowska, E., E.C. Malthouse, and S.F. Bernritter. 2017. Too good to be true: The role of online reviews' features in probability to buy. *International Journal of Advertising* 36, no. 1: 142–63.

Moon, S.J., J.P. Costello, and D.M. Koo. 2017. The impact of consumer confusion from eco-labels on negative WOM, distrust, and dissatisfaction. *International Journal of Advertising* 36, no. 2: 246–71.

Okazaki, S. 2009. Social influence model and electronic word of mouth: PC versus mobile Internet. *International Journal of Advertising* 28, no. 3: 439–72.

Phelps, J.E., R. Lewis, L. Mobilio, D. Perry, and N. Raman. 2004. Viral marketing or electronic word-of-mouth advertising: Examining consumer responses and motivations to pass along email. *Journal of Advertising Research* 44, no. 4: 333–48.

Porter, L. and G.J. Golan. 2006. From subservient chickens to brawny men: A comparison of viral advertising to television advertising. *Journal of Interactive Advertising* 6, no. 2: 30–8.

Precourt, G. 2015. How word of mouth works in advertising. *Journal of Advertising Research* 55 (March): 2–3.

Prendergast, G., D. Ko, and S.Y.V. Yuen. 2010. Online word of mouth and consumer purchase intentions. *International Journal of Advertising* 29, no. 5: 687–708.

San José-Cabezudo, R. and C. Camarero-Izquierdo. 2012. Determinants of opening-forwarding e-mail messages. *Journal of Advertising* 41, no. 2: 97–112.

See-To, E.W.K. and K.K.W. Ho. 2014. Value co-creation and purchase intention in social network sites: The role of electronic word-of-mouth and trust–A theoretical analysis. *Computers in Human Behavior* 31, no. 1: 182–9.

Senecal, S. and J. Nantel. 2004. The influence of online product recommendations on consumers' online choices. *Journal of Retailing* 80, no. 1: 159–69.

Shan, Y. and K.W. King. 2015. The effects of interpersonal tie strength and subjective norms on consumers' brand-related eWOM referral intentions. *Journal of Interactive Advertising* 15, no. 1: 16–27.

Shi, M. and A.C. Wojnicki. 2014. Money talks … to online opinion leaders: What motivates opinion leaders to make social-network referrals? *Journal of Advertising Research* 54, no. 1: 81–91.

Smith, T., J.R. Coyle, E. Lightfoot, and A. Scott. 2007. Reconsidering models of influence: The relationship between consumer social networks and word-of-mouth effectiveness. *Journal of Advertising Research* 47, no. 4: 387–97.

Statista 2017. Virtual reality (VR): Statistics & facts. https://www.statista.com/topics/2532/virtual-reality-vr/ (accessed December 16, 2017).

Steuer, J. 1992. Defining virtual reality: Dimensions determining telepresence. *Journal of Communication* 42, no. 4: 73–93.

Steyer, A., R. Garcia-Bardidia, and P. Quester. 2006. Online discussion groups as social networks: An empirical investigation of word-of-mouth on the Internet. *Journal of Interactive Advertising* 6, no. 2: 51–9.

Strutton, D., D.G. Taylor, and K. Thompson. 2011. Investigating generational differences in e-WOM behaviours: For advertising purposes, does X = Y?. *International Journal of Advertising* 30, no. 4: 559–86.

Taylor, C.R. 2017a. How to avoid marketing disasters: Back to the basic communications model, but with some updates illustrating the importance of e-word-of-mouth research. *International Journal of Advertising* 36, no. 4: 515–9.

Taylor, C.R. 2017b. Native advertising: The black sheep of the marketing family. *International Journal of Advertising* 36, no. 2: 207–9.

Thorson, K.S. and S. Rodgers. 2006. Relationships between blogs as eWOM and interactivity, perceived interactivity, and parasocial interaction. *Journal of Interactive Advertising* 6, no. 2: 39–50.

Vaish, R., Q.V. Liao, and V. Bellotti. 2018. What's in it for me? Self-serving versus other-oriented framing in messages advocating use of prosocial peer-to-peer services. *International Journal of Human Computer Studies* 109: 1–12.

Wallace, E., I. Buil, L. de Chernatony, and M. Hogan. 2014. Who likes you and why? A typology of Facebook fans: From "fan"-atics and self-expressives to utilitarians and authentics. *Journal of Advertising Research* 54, no. 1: 92–109.

Wang, S., N.R. Cunningham, and M.S. Eastin. 2015. The impact of eWOM message characteristics on the perceived effectiveness of online consumer reviews. *Journal of Interactive Advertising* 15, no. 2: 151–9.

Wojdynski, B.W., H. Bang, K. Keib, B.N. Jefferson, D. Choi, and J.L. Malson. 2017. Building a better native advertising disclosure. *Journal of Interactive Advertising* 17, 1–12 doi: 10.1080/15252019.2017.1370401

Wojdynski, B.W. and N.J. Evans. 2016. Going native: Effects of disclosure position and language on the recognition and evaluation of online native advertising. *Journal of Advertising* 45, no. 2: 157–68.

WOMMA. 2017. WOMMA code of ethics and standards of conduct. *Word of Mouth Marketing Association*. http://womma.org/ethics/ (accessed September 6, 2017).

Yim, M.Y., S.C. Chu, and P. Sauer. 2017. Is augmented reality technology an effective tool for e-commerce? An interactivity and vividness perspective. *Journal of Interactive Marketing* 39: 89–103.

Yoon, Y., C. Polpanumas, and Y.J. Park. 2017. The impact of word of mouth via Twitter on moviegoers' decisions and film revenues: Revisiting prospect theory: How WOM about movies drives loss-aversion and reference-dependence behaviors. *Journal of Advertising Research* 57, no. 2: 144–58.

#Me and brands: understanding brand-selfie posters on social media

Yongjun Sung, Eunice Kim and Sejung Marina Choi

ABSTRACT

Marketing scholars and consumer psychologists are turning their attention to the increasing proliferation of selfies. This study investigates a new type of electronic word-of-mouth, namely, selfies with brands/products ('brand-selfies'), on social networking sites, and considers three factors as predictive of brand-selfie posting behaviour: narcissism, materialism, and beliefs that social networking sites are sources of brand information. Data from an online survey were analysed using discriminant analysis to identify characteristics of consumers that do or do not post brand-selfies. The results found that narcissism, materialism, and the belief that social networking sites were brand/product information sources meaningfully related to social networking sites' users' brand-selfie posting behaviour and that they differentiated between brand-selfie posters and non-brand-selfie posters. Consumers' perceptions of social networking sites as sources of brand/product information were most strongly predicted by their brand-selfie posting behaviours. Areas for future research are discussed.

Introduction

Digital technology has given rise to a plethora of venues for consumer self-expressions. Most notable in recent years has been the widespread emergence of social media where huge numbers of consumers reveal their identities. The proliferation of social networking sites (SNSs), propelled by the development of mobile technologies, has allowed consumers to present, faster and more easily than ever before, their personalities, tastes, lifestyles, and favourite brands (Belk 2013). From personal websites and blogs to diverse SNSs, consumers use digital media to present themselves. Perhaps one of the most effective ways to express one's sense of self in digital environments is the 'selfie'.

A 'selfie', named by Oxford Dictionaries as the 2013 word of the year, is 'a photograph that one has taken of oneself, typically one taken with a smartphone or webcam and shared via social media' (Oxford Dictionaries 2013). Millions of selfies are taken every day and posted on a variety of SNSs all over the world, providing individuals with opportunities to show multiple facets of the self (Bazarova et al. 2013; Qiu et al. 2015). A selfie can reflect its producer's true personality (i.e. actual self), but others are somewhat manipulated (i.e. ideal or desirable self-images) because individuals have some extent of freedom to control their features' visibility, their emotional expressions, and camera position compared to other types of photos (Qiu et al. 2015). In addition, a selfie can clearly communicate an individual's passion or interest that reinforces her or his social identity and serve as an artistic expression of that individual's fashion, beauty, and/or possessions (i.e. commercial brands).

As selfies proliferated, the phenomenon attracted the attention of marketing academics and practitioners. Some brands have jumped onto the selfie bandwagon, attracted to the huge potential benefits from including selfies in their overall branding strategies. Perhaps the most popularly established example is Ellen DeGeneres's 2014 celebrity-packed Oscar selfie, which, it was later learned, had been sponsored by the Samsung Galaxy smartphone (Vranica 2014). Despite the selfie's potential to significantly influence consumer culture and marketing practices, it has received limited research attention from marketing scholars and consumer psychologists. Several studies in psychology and communication have investigated personality traits, such as the Five-Factor Model (Ryan and Xenos 2011), narcissism (Fox and Rooney 2015), and jealousy (Christofides, Muise, and Desmarais 2009), as predictors of social media activity and self-presentation behaviour on SNSs. However, few studies have investigated the relatively new phenomenon of the selfie from the perspective of the social media marketplace.

The ways that the selfie could succeed on SNSs as a new type of electronic word-of-mouth (eWOM) depends, to a large extent, on identifying the factors that predict consumers' methods of expressing themselves to influence peer consumers' attitudes and purchasing decisions and how they engage in brand-related eWOM via posting their selfies. For example, some consumers post selfies with brands/products they own (a 'brand-selfie') with a brand-related hashtag (e.g. your #brand). Previous research suggests that, among consumers who follow brands on Twitter, those who re-tweet brand messages outscore those who do not on brand identification, brand trust, community commitment, and community membership intention. These consumers are relatively more likely to be committed to particular brands and to maintain long-term relationships with them (Kim, Sung, and Kang 2014). Similarly, it is reasonable that consumers who take and post brand-selfies are more likely to be committed to those brands and to have longer relationships with them than those who do not post brand-selfies. Thus, brand-selfies are a powerful means of non-verbal communication that link consumers to brands. Of particular interest is that consumers voluntarily associate themselves with brands through their brand-selfies and, thereby, deliberately and immediately express their identities in the public sphere. It would be enlightening to identify who is engaging in this interesting act of posting brand-selfies on SNSs.

This study aims to identify the key factors that lead consumers to post brand-selfies and engage in eWOM on SNSs. We assessed differences in these factors between those who post brand-selfies and those who do not. Specifically, this study examines whether brand-selfie posting behaviour is influenced by two individual-level factors (i.e. narcissism and materialism) and whether brand-selfie posting behaviour relates to consumers' beliefs that SNSs are sources of brand/product information. Therefore, the purpose of the present study is two-fold: (1) to understand the nature of brand-selfie behaviour on SNSs and (2) to identify factors that facilitate, or function to stimulate, consumers' selfie behaviour on

SNSs. This study also examines the relative importance of these factors. We categorized consumers who post selfies into two groups: brand-selfie posters and non-brand-selfie posters, and we examined the factors that distinguish these groups from each other.

The study's findings enhance our theoretical understanding of selfies as an extension of consumers' self-concepts (Belk 1988; Sirgy 1982) and as a powerful tool of eWOM in the rapidly growing world of photo-based social platforms. Besides their theoretical value, the results of this study provide meaningful implications for marketing and advertising managers. Because brands are increasingly personified to respond to consumers' desires for belonging, and because consumers' ownership of certain brands might serve to craft, affirm, and manage their self-identities (Escalas and Bettman 2003), this study helps marketers and advertisers to understand ways that selfies could be leveraged into their marketing communication campaigns and their strategic development of consumer-brand relationships.

Conceptual background and hypotheses development

The selfie as a new type of self-expression

Because consumers are gaining more control over the products and brands they consume, user-generated content is now being considered one of the most critical information sources that help consumers make purchasing decisions. Most notable in recent years is the widespread emergence of eWOM in social media. The proliferation of SNSs, propelled by the development of mobile technologies (e.g. smartphones), has allowed consumers to share, more quickly and easily than ever before, information about products, brands, and their consumer experiences. In addition, they yield a wealth of eWOM in numerous social media environments, where consumers engage in interactions with brands and other consumers through unique SNS applications (e.g. 'likes' and 'shares') and self-generated content (e.g. posts, photos, and comments). Consumers' voluntary diffusion and exchange of brand information is important to understanding the value of SNSs as ideal venues for brand eWOM (Chu and Kim 2011; Kwon et al. 2014).

Recently, photo-based SNS platforms, such as photo-messaging Snapchat, photo-sharing Instagram, and all-in-one Facebook, have provided a key means of self-expression. About 1.8 billion photos are reportedly uploaded and shared daily on a variety of SNSs, such as Flickr, Snapchat, Instagram, Facebook, and WhatsApp (Meeker 2014). The number of selfie uploads is approximately 30% of all uploads among 18–24-year olds (Bennett 2014).

The first selfie hashtag was posted on Instagram in 2011; since then, the selfie phenomenon has continued to grow. Selfie hashtags increased an astonishing 17,000% between 2012 and 2013 (Laird 2013). A recent study by *TIME* analysed Instagram photos tagged selfies from around the world, and identified the top 100 'selfiest cities in the world' (Wilson 2014). The 'Selfie Capital of the World', with approximately 258 active selfie posters per 100,000 people, was Makati, Philippines, followed by Manhattan, New York. Clearly, the selfie phenomenon is not limited to the West, but is quickly becoming a social phenomenon around the world via many types of selfies. Popular types of selfies are labelled with special collaborative terms, such as 'groupie', which refers to a selfie taken with a group; 'foodie', which is a photo of an appetizing meal or food; and 'petfie', which is a selfie taken with a pet.

In the current SNS environments, selfies have become an effective means of self-expression and self-presentation. A recent selfie term is 'braggie' ('bragging' plus 'selfie'), which are selfies shared to show off one's wealth, looks, or social popularity. Typical braggies are photos taken in front of a mirror to show off one's physique next to a cocktail. That selfies are representative of consumers' personalities, interests, and lifestyles that might contribute to facilitating eWOM on SNSs is of particular interest to marketers and advertisers. For example, the 'Outfit of the Day' selfies are most often posted by trendy fashion gurus, who tend to tag the specific brands in their selfies to share brand information.

The brand-selfie

Perhaps one of the newest and most effective ways to communicate brand information and experiences on SNSs is the brand-selfie. Consumers seem to enjoy taking and posting selfies with their brands and products. Consumers intentionally display brand logos or actual products in their selfies for numerous reasons, such as expression of the true or ideal self, social status, or wealth. Regardless of the reasons, those who post brand-selfies aim to associate themselves with those brands, and they expect to benefit from such associations. The act of posting brand-selfies might be driven by a self-presentation motive in the same way that consumers use possessions, brands, or other symbols to construct their images in offline contexts (Belk 1988). Particularly in today's image-focused SNSs, brands are a powerful means of self-expression. Brands generally denote a bundle of meanings, and consumers' ties to brands as ready-made symbols can be depicted and publicized in a controlled way via brand-selfies on SNSs. Some consumers with high levels of brand commitment and self-brand connections might post brand-selfies as public displays of brand support, a public defence of a brand, or as a recommendation to others.

In addition to constructing self-concepts, the use or selection of products or brands serves as a mechanism by which consumers communicate and express themselves to others because individuals make inferences about the values and characteristics of a person based on their possessions (Belk 1988). Early WOM studies primarily focused on the motives and promotional techniques used to spread WOM to influence others' decision-making by highlighting the roles in which consumers want others to view them (e.g. opinion leaders and market mavens). However, it is widely recognized that consumers want to communicate their consumer activities via WOM to express their self-concepts and attract attention (Saenger, Thomas, and Johnson 2013). These consumers might not necessarily be loyal to the brands with which they engage through WOM communication, but they use possessions, products, and brands to transfer meaning from those items to their personal identities (McCraken 1986) and to fulfil a desire to express their self-identities to others. Public displays of consumer activities are considered an effective means of self-expression because they increase the salience of the WOM to other consumers (Taylor, Strutton, and Thompson 2012).

Marketers have sought to tap into the burgeoning selfie phenomenon by incorporating selfies into their marketing strategies. For example, Turkish Airline's 'Kobe vs. Messi: The Selfie Shootout' is an advertisement featuring two iconic sports stars posing for selfies at many of Turkish Air's most popular destinations. With more than 140 million views on YouTube, it was voted YouTube users' favourite advertisement of the decade (Karp 2015). Marketers have tried to encourage user participation by allowing them to post their personal selfies on SNSs and to share them using targeted brand hashtags as a way to build community around a brand and facilitate brand eWOM. For example, to promote its newly launched product, Lancôme introduced the #bareselfie project on its Instagram account; customers could post pictures of themselves without makeup using the #bareselfie tag (King 2014).

In sum, brand-selfies display relationships between consumers and brands and warrant careful examination to learn about consumers' voluntary associations with brands. A recent study on brand-selfies found that, through brand-selfies, consumers extend their brand experiences to their SNSs (Presi, Maehle, and Kleppe 2016). Although similar to eWOM on online platforms, eWOM communications via brand-selfies might reflect consumers' intentions to construct and convey ideal self-identities (Mehdizadeh 2010). The extent to which consumers are motivated to express their personal values and self-identities could be an individual characteristic that varies and that could apply to the display of brand-selfies on SNSs. Previous studies have found that individuals' needs to differentiate themselves from other vary, such as the need for uniqueness (Tian, Bearden, and Hunter 2001), the extent to which they express their identities, namely, identity expressiveness (Thorbjørnsen, Pedersen, and Nysveen 2007), and the need to manage public self-impressions (i.e. self-monitoring; Hall and Pennington 2013). Because brand-selfies might add expressive meanings and value to the narratives that consumers communicate to their audiences (Presi et al. 2016), some individuals could be more motivated than others to produce and post brand-selfies on SNSs. Therefore, in this study, we categorized those who post selfies into two groups: 'brand-selfie posters' and 'non-brand-selfie posters'. Importantly, we examined the factors that distinguish the groups from each other. The following sections discuss the key predictors of brand-selfie posting.

Predictors of brand-selfie posting on SNSs

Narcissism

During the past several decades, social psychologists and personality theorists have devoted significant attention to narcissism (Brown, Budzek, and Tamborski 2009). Recently, a growing body of research on personality has examined narcissism as a predictor of self-promotional behaviours on SNSs (Carpenter 2012; Moon et al. 2016). Narcissistic individuals often hold exaggerated positive self-perceptions, particularly of their physical appearance, social popularity, and status (Bradlee and Emmons 1992; Campbell, Rudich, and Sedikides 2002). Grandiosity is central to narcissism (Buss and Chiodo 1991), along with high extroversion and low agreeableness, a sense of communion, and a need for intimacy (Carroll 1987; Rhodewalt and Morf 1995). SNSs are ideal outlets for narcissists who use interpersonal relationships for self-enhancement and promotion to manage their self-impressions (Wallace and Baumeister 2002). Empirical research on the effects of personality on SNSs' use has found that narcissists are relatively more active in updating their status, posting self-promotional content (e.g. selfies), and acquiring large numbers of 'friends', most often with the goal of displaying inflated social popularity (Fox and Rooney 2015; Ong et al. 2011).

Furthermore, some scholars have characterized narcissists as arrogant power-seekers who want to be unique and admired by others (Carroll 1987; Emmons 1984). Unsurprisingly, narcissism has been found positively correlated with strong desires for wealth (Kasser and Ryan 1996) and material possessions (Rose 2007). Vazire et al. (2008) demonstrated that observers' assessments of narcissism correlated with certain cues, such as expensive and highly fashionable clothing, suggesting that narcissism manifests in physical appearances that could suggest to others a high socioeconomic status. In sum, for narcissists, displaying their material possessions is a way to distinguish themselves from others and be regarded as generally capable and powerful (Christpopher and Schlenker 2000). Based on the above discussion of narcissists' self-promotional behaviours on SNSs, and their tendency to display their material possessions to enhance their shiny self-views, we hypothesized the following.

H1. Brand-selfie posters have higher narcissism scores than non-brand-selfie posters.

Materialism

Belk (1984, 291) defined materialism as 'the importance a consumer attaches to worldly possessions'. In a materialistic society, consumers are surrounded by material objects, which influence them to accept the symbolic meanings designated by brands and/or to which they assign their personal symbolic meanings (Sung and Choi 2012). These symbolic meanings are integral to individuals' expressions of self-concept, lifestyle, socioeconomic status, and to understanding those of others (Solomon 1983). For example, material objects, such as luxury brands, could be used to define consumers personally and their relationships to others and the social environment (Elliot 1997; O'Shaughnessy and O'Shaughnessy 2002; Solomon 1983). As such, the presentation of branded objects in selfies could be understood as a quasi-language through which consumers communicate information about their possessions and social relationships.

Materialists tend to locate material possessions centrally in their lives and to judge their own and others' success by the quantity and quality of their possessions (O'Cass 2004; Richins and Dawson 1992). They tend to value the meanings ascribed to possessions and goods, particularly their social meanings. They believe that ownership of certain products and brands communicates an owner's status, group membership, and other associations. Previous research has suggested that, for individuals who value materialistic goods, pleasure is derived from their communicative aspects once they are acquired. In other words, pleasure relates less to the ownership of the goods than to the positive impressions such ownership communicates to others (Richins 2004). Furthermore, materialists feel satisfied when they share product/brand knowledge and information with other consumers (Fitzmaurice and Comegys 2006). It follows that materialists are relatively more likely to feel pleasure and satisfaction from sharing their product/brand information with others by acting as opinion leaders because materialists are attuned to the social meanings of commercial goods (Fitzmaurice and Comegys 2006). Thus, this study predicted that brand-selfie posters are more likely than non-brand-selfie posters to display high levels of materialism. The following hypothesis formally states this expectation.

H2. Brand-selfie posters have higher materialism scores than non-brand-selfie posters.

Using social media to share brand information

In addition to being a platform for self-expression, SNSs serve as a place for information exchange. Consumers take on multiple roles and actively participate in the consumption, production, and distribution of SNSs' content. SNSs are currently an essential source of information about products and brands. Consumers are increasingly turning away from traditional information sources, where corporate-sponsored messages and advertising prevail, and depending more on SNSs for brand-related information (Vollmer and Precourt 2008). Consumers learn about new brands and products, gather detailed information, and seek out, via SNSs, other consumers' opinions regarding brands before they make purchase decisions (eMarketer 2013). To consumers, SNSs are a more credible and useful source of information concerning brands and purchases because peer consumers have no personal interests in selling products and consumers believe that they share fair information and honest opinions (Foux 2006).

Muntinga, Moorman, and Smit (2011) found that information was an important motivation for consumers using brand-related content on SNSs. Findings from their study suggest that brand-associated content on SNSs gratifies four sub-motivations regarding information: surveillance, knowledge, pre-purchase information, and inspiration. Among them, surveillance and inspiration seem to be most relevant to brand-selfies. By observing brand-selfies, consumers discern new trends and popular fashions, and they can stay abreast of the social environment regarding brands. In addition, others' brand-selfies are sources of motivation. By looking at brand-selfies, consumers often learn about the brands they want to purchase or the brand-related activities (e.g. use occasions, product reforms, fashion styling) in which they want to engage.

The spirit of SNSs is social interaction. Although consumers post brand-selfies for personal identity purposes, these acts are performed with the expectation of social interaction. Consumers knowingly post brand-selfies with the hope that others will recognize and learn from them. If SNSs are perceived to have strong value as platforms for sharing brand information, brand-selfies also might be believed to be powerful and effective. As discussed above, SNSs are increasingly perceived as reliable sources of brand information, and this perception is expected to be stronger among those who post than among those who do not post brand-selfies. This idea is formally stated in the following hypothesis.

H3. Brand-selfie posters perceive greater value in SNSs as brand information sources than do non-brand-selfie posters.

Last, the present study intends to determine the characteristics of brand-selfie posters compared to non-brand-selfie posters. This study empirically investigated whether the three major factors described above significantly distinguish between two groups of selfie posters by brand-selfie posting. Furthermore, the relative strengths of the three factors for predicting brand-selfie posting behaviours were assessed. Thus, the following research questions were posited.

(RQ1) Are brand-selfie posters and non-brand-selfie posters different by the following factors: narcissism, materialism, and SNS value perception?
(RQ2) Which of the factors named in RQ1 best predicts brand-selfie posting behaviours?

Methods

Sample

An online consumer survey was conducted during one week in May of 2015. Respondents were recruited from an online panel managed by the major research firm, Macromill Embrian, in Korea. All of the respondents were volunteer members of a research panel and were notified of the opportunity to participate in this study through e-mail. The online panel comprised more than 1.2 million individuals aged 13 and older (age 13–19 years = 8%, 20–29 years = 38%, 30–39 years = 29%, and 40 years or older = 26%). The research firm offered all of the respondents a virtual currency incentive in return for their participation in the survey.

The initial sample consisted of 319 respondents with experience taking and posting selfies on SNSs. The final sample size (n = 305) reflects the loss of cases due to incomplete questionnaires. Among the 305 respondents, 91 were male and 214 were female. The average age was approximately 29 years (SD = 5.26).

The respondents were asked questions about their selfie behaviours and whether they had ever posted brand-selfies on SNSs. They were then asked to answer questions on narcissism, materialistic orientation, and beliefs about SNSs as sources of brand/product information.

Measures

Narcissism (M = 3.70, SD = .93; α = .89) was measured using a 13-item Narcissism Personality Inventory (Gentile et al. 2013). Example items were: 'I know that I am a good person because everyone keeps telling me so' and 'I will usually show off if I get the chance'. To measure materialism, an adapted 15-item version of Richins' Material Values Scale was used (MVS; 2004; M = 4.06, SD = .79; α = .82); example items are: 'Some of the most important achievements in life include acquiring material possessions' and 'I'd be happier if I could afford to buy more things'. Furthermore, the extent to which respondents valued SNSs as brand/product information sources was measured using two items: (1) 'I believe brand/product information available on SNSs is credible' and (2) I believe brand/product information available on SNSs is useful (M = 4.31, SD = 1.21; α = .82). All of the items were measured using seven-point Likert-type response scales (1 = 'strongly disagree' to 7 = 'strongly agree'). Last, a series of multiple-choice questions asked the respondents to indicate all of the types of selfies that they usually posted on SNSs (e.g. 'daily', 'fashion/beauty', 'brand/product', 'food', and 'party/social').

Results

Preliminary analysis: selfie-posting behaviours

Before testing the hypotheses, descriptive statistics were computed to examine the respondents' selfie behaviours. The average of selfie uploads was about 35.1% of the total photo uploads. The respondents' most preferred SNSs for posting their selfies were Facebook (n = 195; 63.9%) and Korean SNS KakaoStory (n = 158; 51.8%), followed by Instagram (n = 87; 28.5%) and Twitter (n = 21; 6.9%). Respondents preferred taking their selfies in public; that is, outdoor (n = 228; 74.8%) or indoor public places (e.g. classrooms or cafés, n = 215; 70.5%) were the most common locations for taking selfies, whereas a lower percentage of respondents reported that they regularly took selfies in more enclosed or private environments (n = 81; 26.6%). Similar results were reported when respondents were asked to further specify their selfie-taking locations: ordinary everyday places (e.g. school or work, n = 226; 74.1%); restaurants, cafés, or pubs (n = 208; 68.2%); tourist destinations (e.g. airports, tourist attractions, $n = 201$; 65.9%); nature (e.g. forests, parks, $n = 197$; 64.6%); cultural-sports complexes ($n = 117$; 38.4%); inside stores ($n = 70$; 23.0%); inside cars ($n = 45$; 14.8%); and at gyms ($n = 24$; 7.9%).

The ranked order of the most common types of selfies posted by the respondents was: everyday/ordinary ($n = 219$; 71.8%), travel ($n = 166$; 54.4%), leisure ($n = 162$; 53.1%), food ($n = 144$; 47.2%), party/social ($n = 126$; 41.3%), fashion products ($n = 68$; 22.3%), beauty products ($n = 57$; 18.7%), pets ($n = 49$; 16.1%), commercial brands/objects ($n = 46$; 15.1%), and fitness ($n = 14$; 4.6%). If they post at least one type of the fashion products, beauty products, or commercial brands/objects selfies, they are grouped into brand-selfie posters. As a result, 38% ($n = 115$) of the respondents who reported that they posted brand-selfies, including fashion products, beauty products, and commercial brands/objects.

Hypothesis testing: characteristics of brand-selfie posters

Hypotheses 1 through 3 predicted that brand-selfie posters would score higher than non-brand-selfie posters on narcissism, materialism, and the value of SNSs as brand/product information sources. The results found that brand-selfie posters had significantly higher scores on narcissism ($M = 4.05$) than non-brand-selfie posters ($M = 3.47$; $F = 25.59$, $p < 0.001$), on materialism ($M = 4.41$ and 3.85, respectively; $F = 39.52$, $p < 0.001$), and on the perception of SNSs' value as brand/product information sources ($M = 4.91$ and 3.94, respectively; $F = 53.98$, $p < 0.001$). Therefore, H1, H2, and H3 were supported.

Discriminant analysis was employed to address the research question to determine which of the three independent variables (narcissism, materialism, and perception of SNSs' value as brand/product information sources) best discriminated between brand-selfie posters and non-brand-selfie posters. The results found that one function was generated and significant (Wilk's lambda $= 0.70$, $\chi^2(3) = 109.37$, $p < 0.001$), indicating that the function of the three predictors significantly differentiated between the two groups. The eigenvalue was 0.44 and the canonical correlation was 0.55. Brand/product selfie versus non-brand/product selfie poster was found to account for 30.5% of function variance. Hair and others (1998, 294) have suggested that loadings with a value of at least 0.30 should be considered 'substantial'. The structure matrix correlation coefficients (Table 1) revealed that perceptions of SNSs' value as brand/product information sources ($r = 0.64$) was most associated with the function, thereby most strongly predicting when consumers take and post brand/product selfies on SNSs. This was followed by materialism (0.55) and narcissism (0.44).

The original classification results found that brand-selfie posters were classified with 71.3% accuracy whereas non-brand-selfie posters were classified with 88.4% accuracy. In

Table 1. Discriminant functions.

Variable	Mean[a]		Structure coefficient
	Selfies with brand/product ($n = 115$)	Selfies without brand/product ($n = 190$)	
Perception of SNS's value	4.91 (1.11)	3.94 (1.13)*	.64
Materialism	4.41 (.77)	3.85 (.74)*	.55
Narcissism	4.05 (.97)	3.47 (.96)*	.44

[a]Standard deviations are in parenthesis.
*Mean difference was significant at .001.

the overall sample, the results yielded a classification accuracy of 82%. Cross-validation derived 81.6% accuracy for the whole sample. Group means of the function indicated that brand-selfie posters had a function mean of 0.85 and non-brand-selfie posters had a mean of -0.51. The function means results were consistent, suggesting that the respondents with higher scores on narcissism, materialism, and perception of SNSs' value as brand/product information sources were likely to be classified as brand-selfie posters. In sum, these results further confirmed the results of the mean comparisons.

Discussion

The selfie is among the most noteworthy phenomena attracting attention in today's social media environment. Selfies are unique because the person taking the photo and the subject of the photo are the same. In that regard, selfies might be the most fitting type of deliberate self-presentation. It is fascinating to marketers and advertisers that consumers voluntarily include brands in their carefully staged and crafted selfies, which are known as brand-selfies. Brand-selfies have become powerful and unique tools that enable consumers to engage in brand-related eWOM in social media because the information exchange involves a high level of voluntary social communication about brands and self-disclosure (Lee, Im, and Taylor 2008). Accordingly, marketers seek to engage consumers in this new type of eWOM by facilitating their consumer-generated content and exchange.

As one of the first studies to examine this new type of eWOM phenomenon in social media (brand-selfies), this study sheds light on the psychological mechanisms of brand-selfie posting behaviours on SNSs. Its findings suggest that narcissism, materialism, and perceptions of SNSs are significant factors that predict brand-selfie posting behaviours. Furthermore, the three factors significantly differentiate between those who post brand-selfies on SNSs from those who do not. Specifically, posters with more narcissism, materialism, and stronger beliefs in SNSs as sources of brand information are relatively more likely to post brand-selfies.

Brands have symbolic properties, and they can deliver images and messages as social signifiers shared among members of a community. Narcissistic and materialistic consumers are relatively more likely to view brands as effective means of self-expression and to take advantage of their ready-made images for identity construction and presentation.

The results of this study contribute to the marketing and consumer behaviour literature on the concept of the self (i.e. Belk 1988; Sirgy 1982). Understanding consumers' self-concepts is of pivotal importance to marketing scholars and practitioners because consumers' self-perceptions influence and drive their attitudes and behaviours regarding brands and products (Sung and Choi 2012). Consumers often buy, use, and own brands and products as part of an effort to create, maintain, and extend their self-concepts (e.g. Aaker 1997; Belk 1988; Escalas and Bettman 2003; Sirgy 1982). Hoping to establish socially desirable identities (i.e. 'hoped-for-possible selves') on SNSs, posters' brand-selfies might involve producing several brand-selfies and then selecting the images they believe would best present them in a desirable light (Zhao, Grasmuck and Martin, 2008). Brand-selfies extend consumers' brand experiences into their SNSs, and consumers use brands to express and, sometimes, elevate their self-concepts through meaning transfer, which, in turn, influences the brand meanings and images in the marketplace (Presi et al. 2016).

SNSs are a social environment in which consumers easily and deliberately display their brand relationships. The effectiveness of brand-selfies for self-presentation is, of course, based on the belief that SNSs are a platform of information exchange and social interaction. Of particular note is this study's finding that the strongest predictor of brand-selfie posting is consumers' perceptions that SNSs are useful and credible sources of brand information. SNSs are effective social venues of continuous streams of self-images and message production with immediate recognition and feedback. Brand-selfie posters intend to capture the capacity of SNSs, the power of which they are well aware.

The findings of this study offer strategic implications for marketers as much as they provide theoretical insights to the growing selfie phenomenon. Because of their popularity and potential, SNSs have been incorporated into a promotional mix and used as an effective marketing platform from which consumers produce, use, and share brand-related content (Mangold and Faulds 2009). The power of SNSs continues to evolve as the variation in their content evolves. The seemingly most private type of postings are selfies, which have become a new venue for brand presence as increasing numbers of image-based SNS platforms gain popularity. Through this new type of brand eWOM, consumers share meanings to communicate their self-identities with others who are not physically present. In the SNSs where self-presentation tends to be anonymous (Zhao, Grasmuck and Martin, 2008), consumers extend their online realities to resemble their real-life offline situations, which enhances the persuasive impact of brand-selfies on other consumers. The selfie phenomenon could add a new dimension to consumer-brand relationships by providing insights into the ways that brands link consumers' real-life experiences through imagery to their digital self-presentations. When brand-selfies are part of a company-created marketing or advertising campaign, it is important that brands include social desirability, authenticity, and actual experience to help consumers consider SNSs as reliable sources of brand communication and consumer information exchange. Furthermore, because consumers are increasingly using SNSs as sources of brand information, the number of social searches using hashtags is increasing and is somewhat replacing traditional searches engines. Because the voluntary nature of production and posting makes brand-selfies credible and fun, marketers could stimulate and motivate the production and distribution of this consumer-driven brand-content by understanding the nature of active contributors.

Limitations and future research

Although this study's findings yielded significant implications, it is not without limitations, several of which could be considered in light of future research directions. First, to investigate brand-selfie posting behaviours on SNSs, the present study focused only on consumers who post brand-selfies compared to those who do not. Given the diversity of SNSs, and the growing popularity of image-based platforms, such as Instagram and Pinterest, which this study found to garner a considerable portion of brand-selfie postings (38%), future studies should conduct comprehensive examinations of brand-selfie behaviours across an array of platforms with respect to, for example, brand-selfie posting frequency or editing behaviour and their predictive factors. A recent study of brand-related content across SNSs found that the quantity and characteristics of consumer-produced brand-related content varied among Facebook, Twitter, and YouTube (Smith, Fischer, and Yong-jian 2012). Consumers might use different strategies for brand-selfie production, posting, and tagging across the various SNS platforms that they simultaneously use. The effectiveness of brand eWOM on consumer judgement may also vary, depending on the SNS platforms to which the brand-selfies are posted (Lee and Youn 2009).

Second, the present study attempted to predict brand-selfie behaviours based on a limited set of factors related to consumers' self-views and concepts (i.e. narcissism and materialism) and their beliefs in SNSs as brand/product information sources. These have been considered to be crucial factors underlying such self-presentation behaviours (e.g. Fox and Rooney 2015; Sorokowski et al. 2015). It is important to note, nevertheless, that through posting personal photos in a public space – the Web – social meaning and intention are manifest in such behaviours, especially in the context of SNSs in which social relations are important. Future research may therefore examine more social factors such as social support or community identification in relation to individuals' brand-selfie postings.

Third, while this study's findings successfully describe the characteristics of brand-selfie posters, not much insight into their motivations for brand-selfie-posting is offered. Future research could investigate the motivations for posting brand-selfies on SNSs, and how those motivations relate to the behaviours (e.g. frequency, amount) and the types of brand-selfies uploaded. In addition to knowing the kinds of brands involved, our understanding of this phenomenon would expand through an examination of how brands are displayed (e.g. central or peripheral). A viable approach to examining the motivations is through understanding the characteristics of those who diffuse brand eWOM as opinion leaders (Araujo, Neijens, and Vliegenthart 2016; Chu and Kim 2011). Another avenue for future research is the user-side of the story. Little is known about how consumers perceive brand-selfies that others post on SNSs, and future studies will help paint a more complete picture of brand-selfie effectiveness.

While considering this study as a milestone in an exploration of the relationship between individual factors and brand-selfie behaviours, we believe that the brand-selfie phenomenon deserves more attention from marketing scholars and practitioners, and therefore, it is our hope that future studies will advance the endeavours of further in-depth analysis in this important area.

Disclosure statement

No potential conflict of interest was reported by the authors.

Funding

This work was supported by the Ministry of Education of the Republic of Korea and the National Research Foundation of Korea [grant number NRF-2016S1A3A2924760].

References

Aaker, J. 1997. Dimensions of brand personality. *Journal of Marketing Research* 34, no. 3: 347–56.

Araujo, T., P. Neijens, and R. Vliegenthart. 2016. Getting the word out on Twitter: The role of influentials, information brokers and strong ties in building word-of-mouth for brands. *International Journal of Advertising* 36, no. 3: 496–513.

Bazarova, N.N., J.G. Taft, Y.H. Choi, and D. Cosley. 2013. Managing impressions and relationships on Facebook: Self-presentational and relational concerns revealed through the analysis of language style. *Journal of Language and Social Psychology* 30, no. 2: 121–41.

Belk, R.W. 1984. Three scales to measure constructs related to materialism: Reliability, validity, and relationships to measures of happiness. *Advances in Consumer Research* 11, no. 1: 291–7.

Belk, R.W. 1988. Possessions and the extended self. *Journal of Consumer Research* 15, no. 2: 139–68.

Belk, R.W. 2013. Extended self in a digital world. *Journal of Consumer Research* 40, no. 3: 477–500.

Bennett, S. 2014. The year of the selfie: Statistics, facts & figures. *Adweek*, March 19. http://www.adweek.com/socialtimes/selfie-statistics-2014/497309 (accessed June 10, 2016).

Bradlee, P.M., and R.A. Emmons. 1992. Locating narcissism within the interpersonal circumplex and the five-factor model. *Personality and Individual Differences* 13, no. 7: 821–30.

Brown, R.P., K. Budzek, and M. Tamborski. 2009. On the meaning and measure of narcissism. *Personality and Social Psychology Bulletin* 35, no. 7: 951–64.

Buss, D.M., and L.M. Chiodo. 1991. Narcissistic acts in everyday life. *Journal of Personality* 59, no. 2: 179–215.

Campbell, W.K., E.A. Rudich, and C. Sedikides. 2002. Narcissism, self-esteem, and the positivity of self-views: Two portraits of self-love. *Personality and Social Psychology Bulletin* 28, no. 3: 358–68.

Carpenter, C.J. 2012. Narcissism on Facebook: Self-promotional and anti-social behavior. *Personality and Individual Differences* 52, no. 4: 482–6.

Carroll, L. 1987. A study of narcissism, affiliation, intimacy, and power motives among students in business administration. *Psychological Reports* 61, no. 2: 355–8.

Christofides, E., A. Muise, and S. Desmarais. 2009. Information disclosure and control on Facebook: Are they two sides of the same coin or two different processes? *CyberPsychology & Behavior* 12, no. 3: 341–5.

Christopher, A.N., and B.R. Schlenker. 2000. The impact of perceived material wealth and perceiver personality on first impressions. *Journal of Economic Psychology* 21, no. 1: 1–19.

Chu, S.C., and Y. Kim. 2011. Determinants of consumer engagement in electronic word-of-mouth (eWOM) in social networking sites. *International Journal of Advertising* 30, no. 1: 47–75.

Elliott, R. 1997. Existential consumption and irrational desire. *European Journal of Marketing* 31, no. 3/4: 285–96.

eMarketer. 2013. Social media weaves its way through customer life process: Social networks facilitate brand discovery, research and connection. *eMarketer*, November 13. http://www.emarketer.com/Article/Social-Media-Weaves-Its-Way-Through-Customer-Life-Process/1010381 (accessed June 10, 2016).

Emmons, R.A. 1984. Factor analysis and construct validity of the narcissistic personality inventory. *Journal of Personality Assessment* 48, no. 3: 291–300.

Escalas, J.E., and J.R. Bettman. 2003. You are what they eat: The influence of reference groups on consumers' connections to brands. *Journal of Consumer Psychology* 13, no. 3: 339–48.

Fitzmaurice, J., and C. Comegys. 2006. Materialism and social consumption. *Journal of Marketing and Practice* 14, no. 4: 287–99.

Foux, G. 2006. Consumer-generated media: Get your customers involved. *Brand Strategy* 8: 38–9.

Fox, J., and M.C. Rooney. 2015. The dark triad and trait self-objectification as predictors of men's use and self-presentation behaviors on social networking sites. *Personality and Individual Differences* 76: 161–5.

Gentile, B., J.D. Miller, B.J. Hoffman, D.E. Reidy, A. Zeichner, and W.K. Campbell. 2013. A test of two brief measures of grandiose narcissism: The narcissistic personality inventory-13 and the narcissistic personality inventory-16. *Psychological Assessment* 25, no. 4: 1120.

Hair, J.F., W.C. Black, B.J. Babin, R.E. Anderson, and R.L. Tatham. 1988. *Multivariate data analysis*. 5th ed. Upper Saddle River, NJ: Pearson Prentice Hall.

Hall, J.A., and N. Pennington. 2013. Self-monitoring, honesty, and cue use on Facebook: The relationship with user extraversion and conscientiousness. *Computers in Human Behavior* 29, no. 4: 1556–64.

Karp, G. 2015. And the best viral ad of the past decade is…. Chicago Tribune, June 4. http://www.chicagotribune.com/business/breaking/ct-turkish-airlines-ad-0605-biz-20150604-story.html (accessed June 10, 2016).

Kasser, T., and R.M. Ryan. 1996. Further examining the American dream: Differential correlates of intrinsic and extrinsic goals. *Personality and Social Psychology Bulletin* 22, no. 3: 280–7.

Kim, E., Y. Sung, and H. Kang. 2014. Brand followers' retweeting behavior on Twitter: How brand relationships influence brand electronic word-of-mouth. *Computers in Human Behavior* 37: 18–25.

King, J. 2014. Beauty marketers turn to Instagram to stoke product awareness. Luxury Daily, March 26. http://www.luxurydaily.com/beauty-marketers-turn-to-instagram-to-stoke-product-awareness/ (accessed June 10, 2016).

Kwon, E-S, E. Kim, Y. Sung, and C. Yun Yoo. 2014. Brand followers: Consumer motivation and attitude towards brand communications on Twitter. *International Journal of Advertising* 33, no. 4: 657–80.

Laird, S. 2013. Behold the first 'selfie' hashtag in Instagram history. Mashable, November 20. http://mashable.com/2013/11/19/first-selfie-hashtag-instagram/ (accessed June 10, 2016).

Lee, D.H., S. Im, and C.R. Taylor. 2008. Voluntary self-disclosure of information on the Internet: A multimethod study of the motivations and consequences of disclosing information on blogs. *Psychology & Marketing* 25, no. 7: 692–710.

Lee, M., and S. Youn. 2009. Electronic word of mouth (eWOM): How eWOM platforms influence consumer product judgement. *International Journal of Advertising* 28, no. 3: 473–99.

Mangold, W.G., and D.J. Faulds. 2009. Social media: The new hybrid element of the promotion mix. *Business Horizons* 52, no. 4: 357–65.

McCracken, G. 1986. Culture and consumption: A theoretical account of the structure and movement of the cultural meaning of consumer goods. *Journal of Consumer Research* 13, no: 1: 71–84.

Meeker, M. 2014. Internet trends 2014. KPCB, May 28. http://kpcb.com/InternetTrends (accessed June 5, 2015).

Mehdizadeh, S. 2010. Self-presentation 2.0: Narcissism and self-esteem on Facebook. *Cyberpsychology, Behavior, and Social Networking* 13, no. 4: 357–64.

Moon, J., E. Lee, J-A. Lee, T.R. Choi, and Y. Sung. 2016. The role of narcissism in self-promotion on Instagram. *Personality and Individual Differences* 101: 22–25. doi:10.1016/j.paid.2016.05.042

Muntinga, D.G., M. Moorman, and E.G. Smit. 2011. Introducing COBRAs: Exploring motivations for brand-related social media use. *International Journal of Advertising* 30, no. 1: 13–46.

O'Cass, A. 2004. Fashion clothing consumption: Antecedents and consequences of fashion clothing involvement. *European Journal of Marketing* 38: 869–82.

Ong, E.Y., R.P. Ang, J.C. Ho, J.C. Lim, D.H. Goh, C.S. Lee, and A.Y. Chua. 2011. Narcissism, extraversion and adolescents' self-presentation on Facebook. *Personality and Individual Differences* 50, no. 2: 180–5.

O'Shaughnessy, J. and N.J. O'Shaughnessy. 2002. Marketing, the consumer society and Hedonism. *European Journal of Marketing* 36, no. 5/6: 524–47.

Oxford Dictionaries. 2013. Selfie. *Oxford Dictionaries*. http://www.oxforddictionaries.com/definition/english/selfie (accessed June 10, 2016).

Presi, C., N. Maehle, and I.A. Kleppe. 2016. Brand selfies: Consumer experiences and marketplace conversations. *European Journal of Marketing* 50, no. 9/10: 1814–34.

Qiu, L., J. Lu, S. Yang, W. Qu, and T. Zhu. 2015. What does your selfie say about you? *Computers in Human Behavior* 52: 443–9.

Rhodewalt, F., and C.C. Morf. 1995. Self and interpersonal correlates of the narcissistic personality inventory: A review and new findings. *Journal of Research in Personality* 29, no. 1: 1–23.

Richins, M.L., and S. Dawson. 1992. A consumer values orientation for materialism and its measurement: Scale development and validation. *Journal of Consumer Research* 19, no. 3: 303–16.

Richins, M.L. 2004. The material values scale: Measurement properties and development of a short form. *Journal of Consumer Research* 31, no. 1: 209–19.

Rose, P. 2007. Mediators of the association between narcissism and compulsive buying: The roles of materialism and impulse control. *Psychology of Addictive Behaviors* 21, no. 4: 576.

Ryan, T., and S. Xenos. 2011. Who uses Facebook? An investigation into the relationship between the big five, shyness, narcissism, loneliness, and Facebook usage. *Computers in Human Behavior* 27, no. 5: 1658–64.

Saenger, C., V.L. Thomas, and J.W. Johnson. 2013. Consumption–focused self–expression word of mouth: A new scale and its role in consumer research. *Psychology & Marketing* 30, no. 11: 959–70.

Sirgy, J.M. 1982. Self-concept in consumer behavior: A critical review. *Journal of Consumer Research* 9, no. 3: 287–300.

Smith, A.N., E. Fischer, and C. Yongjian. 2012. How does brand-related user-generated content differ across YouTube, Facebook, and Twitter? *Journal of Interactive Marketing* 26, no. 2: 102–13.

Solomon, M. 1983. The role of products as social stimuli: A symbolic interactionism perspective. *Journal of Consumer Research* 10, no. 3: 319–29.

Sorokowski, P., A. Sorokowska, A. Oleszkiewicz, T. Frackowiak, A. Huk, and K. Pisanski. 2015. Selfie posting behaviors are associated with narcissism among men. *Personality and Individual Differences* 85: 123–7.

Sung, Y., and S.M. Choi. 2012. The influence of self-construal on self-brand congruity in the United States and Korea. *Journal of Cross-Cultural Psychology* 43, no. 1: 151–66.

Taylor, D.G., D. Strutton, and K. Thompson. 2012. Self-enhancement as a motivation for sharing online advertising. *Journal of Interactive Advertising* 12, no. 2: 13–28.

Thorbjørnsen, H., P.E. Pedersen, and H. Nysveen. 2007. "This is who i am": Identity expressiveness and the theory of planned behavior. *Psychology & Marketing* 24, no. 9: 763–85.

Tian, K.T., W.O. Bearden, and G.L. Hunter. 2001. Consumers' need for uniqueness: Scale development and validation. *Journal of Consumer Research* 28, no. 1: 50–66.

Vazire, S., L.P. Naumann, P.J. Rentfrow, and S.D. Gosling. 2008. Portrait of a narcissist: Manifestations of narcissism in physical appearance. *Journal of Research in Personality* 42, no. 6: 1439–47.

Vollmer, C., and G. Precourt. 2008. *Always on: Advertising, marketing, and media in an era of consumer control*. 1st ed. New York, NY: McGraw-Hill.

Vranica, S. 2014. Behind the preplanned Oscar selfie: Samsung's ad strategy. *The Wall Street Journal*, March 3. http://www.wsj.com/articles/SB10001424052702304585004579417533278962674 (accessed June 10, 2016).

Wallace, H.M., and R.F. Baumeister. 2002. The performance of narcissists rises and falls with perceived opportunity for glory. *Journal of Personality and Social Psychology* 82, no. 5: 819.

Wilson, C. 2014. The definitive ranking of the selfiest cities in the world. *Time*, March 10. http://time.com/selfies-cities-world-rankings/(accessed June 10, 2016).

Zhao, S., S. Grasmuck, and J. Martin. 2008. Identity construction on Facebook: Digital empowerment in anchored relationships. *Computers in human behavior* 24, no. 5: 1816–36.

Understanding the effects of different review features on purchase probability

Su Jung Kim [iD], Ewa Maslowska and Edward C. Malthouse

ABSTRACT

The role of electronic word-of-mouth (eWOM) has been recognized by marketers and academics, but little research has examined the impact of eWOM on purchase behavior. Building on dual-process models of persuasion, this study aims to disentangle the effect of different online review features (i.e. argument quality, review valence, review helpfulness, message sidedness, source credibility and reviewer recommendation). Using product reviews and purchase data from an online retailer website, we investigate the financial impact of online product reviews on purchase decisions. The results demonstrate the persuasive power of different review features that are derived from dual-process models of information processing. Managerial implications on how advertisers and companies should design and manage online product reviews are offered.

With the emergence of digital and social media, electronic word-of-mouth (eWOM) has become a powerful source of information influencing purchase decisions. Consumers have constant access to online product reviews from online retailers, brand websites, brand community blogs, and third-party review platforms where consumers can participate and engage in discussions about their consumption experience. Among various forms of eWOM, this study focuses on online product reviews because they are written by consumers who presumably have experience with a product and are actively sought by potential consumers, thereby affecting review readers' purchase decisions more directly. A growing number of studies have identified various effects of online product reviews on consumer attitudes and behaviors (Chevalier and Mayzlin 2006; Liu 2006; Maslowska, Malthouse, and Bernritter 2017; Maslowska, Malthouse, and Viswanathan 2017). A recent report from the Nielsen Company (2015) finds that consumers trust recommendations or opinions from other consumers more than traditional forms of advertising such as commercials or product placements on mass media, showing the persuasive power of online product reviews.

Several features of online product reviews have been a subject of scholarly investigation including review quality, valence, mode (e.g. textual, visual, or multi-modal), platform (e.g. company-provided vs. third-party provided), or reviewer characteristics (e.g. experts vs. non-experts). Many studies have examined how the aforementioned online review features influence the level of usefulness or helpfulness of reviews. Some have investigated how a single or combination of these features affect psychological variables such as brand attitudes or trust, or behavioral variables such as purchase intention (Dhanasobhon et al. 2007; Doh and Hwang 2009; Floyd et al. 2014).

The purpose of this study is as follows: first, it extends previous scholarly efforts on understanding the influence of online product reviews and examines the effects of different review features by testing multiple review characteristics in a single model. In doing so, we also test a possible curvilinear relationship and interaction effects of review features to advance our understanding of the relationship among review characteristics.

Second, this study estimates the monetary impact of online product reviews by linking individual consumers' exposure to reviews with their actual purchases. The majority of previous studies have focused on psychological variables such as attitudes (see Purnawirawan et al. 2015, for a review) or used proxy measures such as sales rank, but research studying actual sales at an individual level is limited (see Floyd et al. 2014, for a review). This study uses online product reviews and individual-level sales data, which enables us to provide empirical evidence of the impact of online product reviews on actual sales. While many existing studies treat review helpfulness (often referred to as usefulness) as a dependent variable (e.g. Mudambi and Schuff 2010; Quaschning, Pandelaere, and Vermeir 2015; Schindler and Bickart 2012; Willemsen et al. 2011), this study considers helpfulness as one of the predictor variables to find out its role in making a purchase decision.

Finally, in contrast to previous research that has focused on experiential goods from Amazon, TripAdvisor or Yelp, this study uses data from one of the largest online retailers, which sells various types of search goods (e.g. household items, beauty products, or over-the-counter medicine). This gives us an opportunity to investigate which elements of reviews (review content and/or non-content cues) would matter more when it comes to purchasing everyday products.

In sum, this study looks at the association of different online product review features (i.e. review valence, length, pros and cons, helpfulness, authorship, and product recommendation) with purchase probabilities. By investigating these features, this study offers theoretical contributions to the literature on information processing as well as managerial insights regarding how advertisers can use reviews and how firms should manage their online recommendation systems to better serve existing and potential consumers.

Literature review

Online product reviews and the dual-process models of persuasion

Online product reviews can be described in terms of quantitative and qualitative features (Sridhar and Srinivasan 2012). Quantitative aspects of reviews are often expressed as numerical summaries such as average star ratings and number of reviews. Qualitative aspects present consumers' assessment of a product or a service such as review content. Because quantitative aspects are often displayed above or next to a product description, they are read by customers without them perusing the textual portion of reviews.

Qualitative aspects, on the other hand, require customers to either click on a 'Review tab' or scroll down a web page in order to find and comprehend a text. Usually qualitative review features require additional attention and/or action from customers compared to quantitative review features.

This distinction of quantitative and qualitative aspects of online product reviews and the differential levels of attention and motivation required to process them lead us to use the dual-process models of persuasion such as the Elaboration Likelihood Model (ELM) (Petty and Cacioppo 1986) or the Heuristic Systematic Model (HSM) (Chaiken 1980) as our conceptual framework to understand the persuasiveness of different online review features. These models are based on the premise that there are two distinctive routes of information processing. Central route or systematic processing assumes that individuals have the ability and motivation to process messages, which results in deeper information processing and lasting attitudinal changes. In contrast, peripheral route or heuristic processing is associated with using simple decision rules and cues, leading to superficial information processing and temporary attitudinal changes.

The ELM has been widely used as a framework to explain the ways in which consumers process information in online product reviews (Cheng and Ho 2015; Cheung, Sia, and Kuan 2012; Park, Lee, and Han 2007). However, this study takes the HSM as an overarching theoretical framework because of its flexibility in terms of applying the dual paths to persuasion (Zhang and Watts 2008). Unlike the ELM, which assumes that individuals take either the central or peripheral route to persuasion, the HSM posits that the heuristic and systematic processing may occur independently or concurrently, allowing us to investigate the simultaneous impact of heuristic and systematic processing (Bohner, Chaiken, and Hunyadi 1994). The HSM specifies three motivations of information processing, namely, accuracy, defense, and impression motivation. Among these three, the accuracy motivation (i.e. the motivation to make objective judgements) is most closely aligned with the context of online product reviews. Todorov, Chaiken, and Henderson (2002) pointed out that both systematic and heuristic processing can lead to accuracy, urging the need to incorporate message or source characteristics related to the two processes. It is noteworthy that the purpose of this paper is not to test the conditions of how systematic and heuristic processes take place. Rather, we apply the dual-process framework of the HSM to discuss the persuasive effects of online product reviews. Similar to the approach taken by Zhang et al. (2014), we consider that both content-related characteristics and non-content cues affect review readers' purchase decisions concurrently.

Online reviews can influence consumers via heuristic or systematic processing since they are composed of content-related characteristics (e.g. argument quality) as well as non-content cues (e.g. average star ratings). Table 1 summarizes extant research that applied the dual-process models to online product reviews. As shown in Table 1, previous research has treated argument quality as an important element in systematic processing. Other content-related or source-related features such as star ratings or source credibility have been regarded as heuristic cues. Consumers who read reviews can be influenced by a single feature or a combination of them. In addition to argument quality, star ratings, and source credibility, this study includes review helpfulness, message sidedness, and reviewer recommendation as other possible cues that consumers can use while they are processing online product reviews. We divide these characteristics into message content and message source (i.e. reviewer) characteristics.

Table 1. Summary of existing literature on eWOM that applied dual-processing models.

Author(s)	Theory	Method	Predictor(s)	Mediator or moderator	Outcome(s)	Findings
Baber et al. (2016)	HSM	Survey	Trustworthiness Expertise Experience WOM use	Attitude	Intention to purchase electronic products	E-WOM sources' levels of trustworthiness and experience positively influenced eWOM use. E-WOM use positively affected attitude, which fully mediated its effect on purchase intentions
Cheng and Ho (2015)	ELM	Secondary analysis of exiting reviews	Argument quality Source credibility		Review usefulness	Argument quality and source credibility all have a significant positive effect on the readers perception of the usefulness of reviews. The effects of source credibility exert a larger influence than argument quality
Cheung et al. (2008)	ELM	Survey	Argument quality Source credibility		Information usefulness Information adoption	Argument quality (reliance, timeliness, accuracy, comprehensiveness) and source credibility (expertise, trustworthiness) both have a positive influence on information usefulness, which in turn has a positive effect on information adoption
Cheung, Sia, and Kuan (2012)	ELM	Survey	Argument quality Source credibility Review consistency Review sidedness	Recipients' expertise Recipients' involvement	Review credibility	Argument quality, source credibility, review consistency, and review sidedness all have a positive effect on review credibility. The effect of review sidedness on review credibility is stronger for recipients with lower levels of involvement
Filieri and McLeay (2013)	ELM	Survey	Information quality Information quantity Product ranking		Information adoption	Some dimensions of information quality (timeliness, relevance, accuracy, value-added information) and product ranking have a positive influence on travelers' adoption of information from online reviews
Gupta and Harris (2010)	HSM	Experiment	E-WOM argument strength Optimality of product choice	Need for cognition	Product choice Total time spent on the site Time spent considering recommended options	E-WOM recommendations on an experience product lead high-NFC individuals to spend significantly more time analyzing their choices than do low-NFC individuals. Low-NFC consumers make suboptimal choices based on e-WOM recommendations, whereas high-NFC consumers tend to use e-WOM recommendations, but follow the recommendation only if it is an optimal product. Consumers, particularly high-NFC consumers, are willing to move away from their existing preferences, given that e-WOM recommendations on an experience product are probably deemed valuable enough for them to sacrifice their own preferences
Kim, Brubaker, and Seo (2015)	Dual-process models	Experiment	Perceived authority Perceived bandwagon Perceived	Credibility	Product attitude Webpage attitude Purchase intention	Expert reviews have a greater impact on attitudes toward product review websites, and this effect was moderated by star ratings. Star ratings have a strong positive effect on users' attitudes toward the product, attitudes toward the

(continued)

Table 1. (*Continued*)

Author(s)	Theory	Method	Predictor(s)	Mediator or moderator	Outcome(s)	Findings
			objectivity Social plugins Star ratings			website, and their purchase intention. Presence of sharing applications also had positive effects on attitudes toward the product. Credibility mediates the relationship between heuristic cues and product evaluation
Park, Lee, and Han (2007)	ELM	Experiment	Review quantity Review quality	Involvement	Purchase intention	The quality and quantity of online reviews positively affect consumers' purchase intention. Low-involvement consumers are affected by the quantity rather than the quality of reviews, whereas high-involvement consumers are affected by review quantity mainly when the review quality is high
Park and Kim (2008)	ELM	Experiment	Expertise of review readers Types of reviews Number of reviews		Purchase intention	The effect of type of reviews (cognitive fit) on purchase intention is stronger for experts than for novices while the effect of the number of reviews on purchase intention is stronger for novices
Sher and Lee (2009)	ELM	Experiment	Argument quality Source credibility	Skepticism	Purchase intention	High-skepticism consumers do not take the central route, but tend to base their attitudes on intrinsic beliefs. Low-skepticism consumers adopt the peripheral route in forming attitudes
Zhang and Watts (2008)	HSM	Survey	Argument quality Source credibility	Level of disconfirming information-focused search	Information adoption	Argument quality and source credibility can affect the adoption of online reviews in online communities. Disconfirming information had a moderating effect only in an online travel forum, not in a community of computational fluid dynamics
Zhang et al. (2014)	HSM	Survey	Argument quality Source credibility Perceived quantity of reviews		Behavioral intention	Argument quality, source credibility, and perceived quantity of reviews were key determinants of behavioral intention. The bias effects from source credibility and perceived quantity of reviews toward argument quality were also found

Review content characteristics

Argument quality

One of the fundamental elements in online reviews is content. People who have the motivation and ability to read reviews (i.e. systematic processing) will pay careful attention to other consumers' opinions about a product that they consider purchasing. People who lack the motivation or ability to read reviews will glance over other cues that signal the quality of a message (i.e. heuristic processing). Previous studies have found that argument quality positively influences information adoption (Zhang and Watts 2008) or purchase intention (Park, Lee, and Han 2007; Zhang et al. 2014). For example, Zhang et al. (2014) found that argument quality (i.e. perceived informativeness and persuasiveness) is one of the key determinants of consumers' willingness to purchase products. Park, Lee, and Han (2007) identified review quality (i.e. relevance, objectiveness, understandability, sufficiency) as one of the antecedents of purchase intention. In many studies, *length* of a message has been regarded as a proxy for information quality (Bosman, Boshoff, and van Rooyen 2013; Cheng and Ho 2015; Huang et al. 2015; Mudambi and Schuff 2010) with longer messages expected to provide high-quality information on product features. Extant studies suggest a positive linear relationship between review length and its effect on review credibility, helpfulness, or purchase intention.

However, it is also possible that the effect of review length is nonlinear. The reasoning behind this is twofold. First, although longer messages may induce greater certainty than shorter ones, because they are perceived as more complete (Rucker et al. 2014), previous research has observed that customers write long reviews to express their dissatisfaction (Vasa et al. 2012). Second, individuals have a limited cognitive capacity and hence cannot attend to and process all available stimuli (Kahneman 1973). Therefore, consumers may not be willing or able to comprehend a message that is too lengthy due to cognitive overload (Huang et al. 2015; Kuan et al. 2015). This implies that the positive impact of review length will reach its maximum at a certain length and then decrease once the length passes this threshold. Thus, we hypothesize that review length has a curvilinear relationship with purchase decision, meaning that the positive effect increases until the length of a review reaches a threshold, and then diminishes.

H1: Review length has an inverted-U relationship with purchase probability.

Review valence

While the length of a message is an indicator of the quality of the message, the *valance* of a review is an indicator that implies the cognitive consequence of customers' attitudes toward a product (Liu 2006). In an online review system, review valence is expressed in a form of star ratings, which serve as a heuristic cue reflecting the popularity and the quality of a product (Sundar, Oeldorf-Hirsch, and Xu 2008). Extant literature is equivocal regarding the effects of ratings (King, Racherla, and Bush 2014), but the majority of previous studies have found that a higher average star rating is associated with more favorable impressions of products, which increases purchase intention (Chen 2008; Kim, Brubaker, and Seo 2015; Sundar, Oeldorf-Hirsch, and Xu 2008). More specifically, a positive effect of star rating on sales (rank) has been found for such products as books, movies, cell phones or beer (Chevalier and Mayzlin 2006; Chintagunta, Gopinath, and Venkataraman 2010; Clemons, Gao, and Hitt 2006; Gopinath, Thomas, and Krishnamurthi 2014; Zhang and Dellarocas 2006). In line with these studies, we expect that positive product reviews improve consumers' attitudes toward products and increase purchase intention, whereas negative ones exert the opposite impact.

Interaction between review valence and helpfulness

In addition to this positive impact of positive reviews and the negative impact of negative reviews, we expect that this association between valence and purchase probability is moderated by the level of review helpfulness. Previous research into online reviews has overwhelmingly focused on predicting helpfulness of online reviews, hence, treating it as an outcome measure. Therefore, we have some understanding of what makes reviews helpful, but we do not know what the role of helpfulness is in the purchase decision process. Scarce previous research suggests that more helpful reviews are more prevalent for top selling products (Otterbacher 2009), and that helpfulness is a signal of other customers' endorsement of the review (Metzger, Flanagin, and Medders 2010), which translates to a more positive attitude towards the reviewer and product (Walther et al. 2012). Extant research also suggests that consumers perceive extreme reviews (i.e. very negative or very positive) more useful than moderately rated reviews (Park and Nicolau 2015). Taken together, this suggests that there will be an interaction between review valence and helpfulness in a way that positive reviews that are perceived more helpful will exert a more positive influence than positive reviews that are perceived less helpful. In case of negative reviews, the pattern will be the opposite: negative reviews with a higher level of helpfulness would have a more negative impact compared to negative reviews with a lower level of helpfulness. Therefore, we pose the following hypothesis that predicts the moderating role of review helpfulness on the association between valence and purchase probability.

H2: The level of review helpfulness moderates the association between valence and purchase probability. In particular, there is a positive interaction effect between valence and the level of review helpfulness on purchase probability.

Review sidedness

While review valence provides an overall evaluation of a product from a negative to positive spectrum, *pros* and *cons* (i.e. a summary of pros and cons of a product) demonstrate whether a reviewer provides a summary of positive and negative aspects of a product. Some companies started asking reviewers to write down specific pros and cons or to choose them from a provided scroll-down menu, believing that the presence of pros and cons can make reviews more persuasive, which would be in line with advertising literature studying the effect of two-sided messages. However, in the context of eWOM, it may not be the case. Schlosser (2011) pointed out that, in the case of online reviews, one-sided arguments can be considered more helpful and persuasive than two-sided ones unlike advertising messages. The aim of advertising is to sell a product and hence is not perceived as credible when one-sided arguments are provided. However, online reviews are not expected to be driven by persuasion motives, but motives to share opinions and help other consumers make an informed purchase decision, which makes one-sided arguments more persuasive. Therefore, we hypothesize that the presence of pros will have a positive influence on purchase probability, whereas the presence of cons will have an opposite influence. Additionally, we ask whether there is an interaction between the presence of pros and cons. In other words, we ask whether two-sided messages (i.e. reviews presenting both pros and cons) have a different impact on purchase probability compared to one-sided messages (i.e. reviews showing only pros or cons).

H3a: The presence of pros in a review is positively associated with purchase probability.
H3b: The presence of cons in a review is negatively associated with purchase probability.
RQ1: Is there an interaction between the presence of pros and cons in a review (i.e. a two-sided review)?

Reviewer characteristics

Source credibility

Source credibility has been recognized as an important element of persuasion. When a person perceives a source to be trustworthy or have the expertise on a topic, it is more likely that a message from the source is seen as more credible (Quaschning, Pandelaere, and Vermeir 2015; Sundar 2008). The trustworthiness and expertise of the source have been extensively studied as major dimensions of source credibility in advertising research (Dou et al. 2012). In addition, as shown in Table 1, source credibility has been identified as a major heuristic cue in the dual-process models. Previous studies have found that reviews written by consumers are perceived as more believable and understandable than those written by experts or companies because they provide users with information based on their actual product experience (Li et al. 2013; Riegner 2007).

However, consumers are aware that some firms can remove negative reviews or encourage positive reviews with some forms of rewards (Li and Hitt 2008). This can make consumers skeptical about the trustworthiness of reviews. To address such concerns, several online retailers, including Amazon.com, place a *verified buyer* badge to indicate whether a reviewer made a purchase. For other consumers, this shows that the reviewer has the experience of using the product, which increases the level of expertise and trustworthiness. In addition, it suggests that the reviewer is a real consumer and not someone who was paid to write a review. Because people are more inclined to trust those similar to ourselves (McCroskey, Richmond, and Daly 1975), consumers are more likely to follow information provided by other customers (Blazevic et al. 2013; Huang et al. 2015). Therefore, we hypothesize that reviews written by customers who are verified customers will have positive effect on purchases.

H4: The presence of a verified buyer badge is positively associated with purchase probability.

Reviewer recommendation

In addition to an indicator of verified purchase, reviews contain *recommendations* from those who create reviews (Park and Kim 2008). Reviewers can indicate whether they would recommend a product to a friend. Unfortunately, previous research into the role of this review feature is scarce. This is surprising since, as Reichheld (2003) claims, a customer's propensity to recommend a product to others (i.e. referral value) is the most important success measure in business. In addition to star ratings, which signal an overall evaluation of product quality, reviewer's recommendation provides a measure of reviewers' willingness to recommend the product and constitutes another cue indicating the reviewer's satisfaction with the product (e.g. Finn, Wang, and Frank 2009). As such, it can directly influence other consumers' purchase intention. We propose the following hypothesis.

H5: The intention of a reviewer to recommend a product to a friend is positively associated with purchase probability.

Method

Data

This study uses online product reviews and purchase data obtained from a large online retailer in the United States. The company provides health, beauty, and personal care items such as over-the-counter medicine, vitamins, cosmetics, and skin/hair products. Since the company only has an online presence, all purchases are recorded and can be linked to product reviews on the firm's website. The company provides an online review page that allows its users to post and read product reviews. Once customers find a product, they can browse the product page with the image, price, star ratings, and number of available reviews for the given item. More interested consumers can click a 'Review' tab positioned below a brief product description and picture.

Once consumers click the Review tab, they can see more detailed review features. First, they see a summary of reviews when there are more than two reviews for a given product. The aggregated information includes average star ratings (from 1 to 5 stars), the number of reviews available for the product, and the percentage of reviewers who said they would recommend the product to a friend. When there is a single review for a product, the summary section is not provided. Below the summary information is each review sorted from the newest to oldest by default. Figure 1 illustrates how each review is shown to the users and what elements are included in each review: (1) The date when the review was written; (2) under the date of post, a star rating is presented on a scale of 1 star from 5 stars; (3) then, the information on the reviewer is provided including the name (can be a real name or nickname) and a badge that shows whether the reviewer is a user who has a record of verified purchase from the retailer; (4) optionally, reviewers can choose whether they list a list of pros, cons, and/or best uses of the product; (5) review text is presented; (6) optionally, reviewers can choose to answer to the question whether they would recommend the product to a friend; (7) finally, review readers can vote whether they think the given review was helpful (Yes/No).

We have information on browsing (click log table) and purchase activities (transaction table) for 14 weeks, starting from 29 June 2014 through 11 October 2014. In addition, we have reviews data that include information on review and reviewer characteristics. However, the data do not allow us to track which specific reviews were read by a customer, the time spent reading reviews, or whether the customer changed the sort order. Rather, we are only able to know whether or not a customer clicked the Review tab for a specific product. Due to this constraint, we narrow down our analysis to products that have a single review to make sure that those who clicked on the Review tab actually were exposed to the review. In sum, a total of 9838 products that have one review are selected for analyses. These products are displayed 420,334 times during the 14-week period. Our unit of

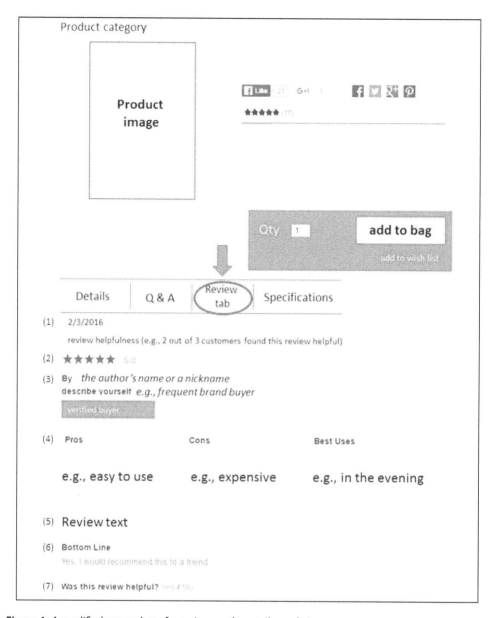

Figure 1. A modified screenshot of a review on the retailer website.

analysis is an exposure (i.e. display) of a product review to a consumer who clicked the Review tab on the retailer's website.

We acknowledge that this analytic approach limits our ability to generalize the findings. However, it gives us the ability to estimate the influence of specific review features on purchase probability more accurately. First, it allows us to link a review with purchases from those who read this specific review because this is the only review available to those who were interested in the product, which helps us better estimate the financial impact of individual reviews. Second, this analytic setting makes us certain that reviewers are taking either systematic or heuristic information processing or both. Since there is a single review for a product, consumers clicking the Review tab will be exposed to both review content and other information cues. Following the accuracy motivation, those who are highly motivated to make an objective judgement about their product choice will go through systematic processing by reading the argument(s) in review content. Those who are less motivated simply scan through heuristic cues such as a reviewer badge or star ratings (Maslowska, Malthouse, and Viswanathan 2017). It is also possible for customers to use both factors so that systematic and heuristic processing occur concurrently.

Measures

Predictor variables

The predictor variables are online review and reviewer characteristics that consumers are exposed to when they process online product reviews systematically and/or heuristically. Regarding review content characteristics, *argument quality* is measured by review length (i.e. number of words in a review). We create a quadratic term of review length to see whether the relationship between review length and purchase probability is an inverted-U shape, as stated in H1. *Review valence* refers to whether a review is negative, neutral, or positive. We categorize reviews into negative, neutral, and positive reviews with three stars as a cut-off value. Reviews with three stars are coded as neutral reviews and reviews below/above three stars as negative/positive reviews. The majority of the reviews were positive (77.5%), followed by negative (15.7%) and neutral ones (6.8%). *Review helpfulness* is operationalized as the level of helpfulness of a review as indicated by other consumers. As shown in Figure 1, at the bottom of each review, review readers can vote whether it was helpful or not. We use the raw count of the response 'Yes' to the question as the measure of review helpfulness. Message sidedness is measured by using pros and cons variables. *Pros* and *cons* are two binary variables showing whether a review has a presence of pros or cons of a product. Thus, if a review has at least one mention of pros, the value becomes 1, otherwise 0. The same applies to cons. Initially, we computed the count of pros and cons in each review. However, because there was a large number of missing values, pros and cons were dichotomized. Out of 420,334 exposures, 16% presented a list of pros, whereas 3% displayed a list of cons.

With regard to reviewer characteristics, we create two predictor variables: source credibility and reviewer recommendation. *Source credibility* is a binary variable indicating whether a reviewer is a 'verified buyer,' meaning that he/she purchased the item on the company's website. Out of 420,334 exposures, about half (52%) show that a review is written by a verified buyer, whereas the other half (48%) fall into a review written by an anonymous reviewer who is not a verified buyer. This does not necessarily mean that they have never purchased the product. It is possible that such reviewers have bought the product elsewhere. *Reviewer recommendation* is a categorical variable that shows whether a reviewer answered 'Yes' to the question, 'Would recommend this to a friend?', as shown in item (6) in Figure 1. About 77% report they would recommend the product to a friend, 15.9% report they would not and 7% do not provide any answer.

Outcome variable

Our outcome variable is whether or not an item is purchased (i.e. conversion) after an exposure. Among the 420,334 exposures we analyzed, about 5.5% converted to purchases. A summary of descriptive statistics of the main variables is presented in Table 2.[1]

Table 2. Summary of descriptive statistics of main variables ($N = 420{,}334$).

	Mean	S.D.
Review content characteristics		
Review length (# of words)	54.64	53.51
Review age (days)	656.91	495.44
Review helpfulness (votes)	1.42	2.98
Product characteristics	0.66	0.47
Price (dollars)	16.63	21.05

Analysis

To test our hypotheses and research question, we estimate a logistic regression predicting the probability of product purchase. We include characteristics of review content and review contributors as predictor variables. We also include review age (i.e. the number of days since the date when the review was written), product price, product category (i.e. broad product category presented on the retailer's website, for example, medicine & health, beauty, household items, baby & mom, etc.), and month of purchase (i.e. seasonality) in the model as control variables. Review age ranges from 2 days to 4249 days (median = 575 days). About one-third (35.1%) of exposures occurred in the beauty section of the website, followed by medicine & health (27.1%), personal care (14.9%), household, food, & pets (7.4%), sexual well-being (6.8%), baby & mom (4.4%), and others (4.2%). Predictor and control variables having a skewed distribution are log-transformed before they are entered into the model.

We discuss the rationale for our model specification. The purpose of the paper is to study the effects of all the review stimuli on the purchase decision of a prospective buyer. We therefore include measures of all stimuli, since this is what the prospective customer sees. There will likely be correlations between some of the variables, because the different features of a review reflect the reviewer's experience with the product. For example, a reviewer who had a positive experience will likely give a large number of stars, be willing to recommend the product to a friend, and list pros, creating positive correlations between the variables. Dropping measures of certain stimuli from the model will create an omitted variable bias and overstate the effects of the remaining variables (Liu-Thompkins and Malthouse 2017). Measures of all customer stimuli must therefore be included in one model.

Results

A logistic regression analysis is conducted to estimate purchase probability using characteristics of review content and reviewers with $n = 413{,}666$. A test of the full model against a constant only model is statistically significant, showing that the independent variables as a set predict purchase behavior ($\chi^2_{25} = 178928.5 - 175207.9 = 3720.638, p < .0001$). Cox and Snell's R-squared equals 0.008954, Nagelkerke's R-squared equals 0.02550, and McFadden's R-squared equals 0.02079. While the primary objective of this model is not prediction,[2] the area under the receiver operating characteristics (ROC) curve (AUC) equals 0.615, with the ROC curve shown in Figure 2.

Table 3 presents the logistic regression results. Before discussing the parameter estimates, we assess and discuss multicollinearity. Multicollinearity is where the predictor

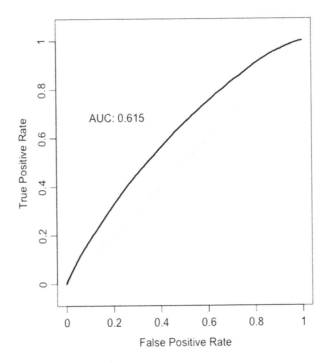

Figure 2. ROC curve of the fitted model.

variables are correlated with each other. The effect of multicollinearity is to increase the standard errors of the estimates, but multicollinearity does not affect other properties of the estimates, for example, they are still unbiased and the standard errors correctly reflect the reduced stability of the estimates. Liu-Thompkins and Malthouse (2017) suggest evaluating the correlations between predictors with variance inflation factor (VIFs), which quantify how much the variance (squared standard errors) of an estimated coefficient increases because of collinearity. When there are categorical predictors or polynomials, as we have here, then generalized VIFs (GVIF) can be used[3] (Fox and Monette 1992). GVIFs are provided in Table 3. The terms with an "X" indicate an interaction, where pros × cons has two dummy main effects and a product for the interaction. The valence × loghelp has two dummies for valence (neutral and positive), a slope for loghelp, and two interaction terms (loghelp × neutral and loghelp × positive), using a total of five degrees of freedom. Many of the GVIF values are substantial, with the largest equaling 6.05 for the valence × help interaction. Standard errors are also a function of sample size, where larger samples produce more reliable (lower standard error) estimates. While there is multicollinearity, inflating standard errors, our large sample sizes combat the variance inflation yielding small standard errors.

Concerning H1 (i.e. argument quality), we found that the association between review length and purchase probability shows an inverted-U shape. However, we only found directional evidence at the .10 level, thus failing to confirm H1. Figure 3 illustrates the quadratic effect of review length on purchase probability. Directionally, this suggests that the effect of review length is positive until it reaches its vertex (between 20 and 55 words) and then becomes negative when a review is too lengthy (i.e. above 55 words), showing

Table 3. Logistic regression predicting purchase probability.

Predictors	B	S.E.	Wald χ^2	GVIF(df)	Odds ratio
Intercept	1.7362***	0.488	12.655		
Review length				1.23(2)	
Log(Review length)	0.145	0.089	2.664		1.156
Log(Review length) squared	−0.021	0.011	3.400		0.979
Review valence					
Neutral	−2.118**	0.725	8.530		0.1203
Positive	−2.408***	0.460	27.437		0.0899
Negative (1–2 stars)	0	–	–		
Review helpfulness	−1.471***	0.188	61.053		0.2297
Review valence × helpfulness				6.05(5)	
Neutral × helpfulness	0.932**	0.305	9.357		2.540
Positive × helpfulness	1.153***	0.192	27.437		3.168
Review sidedness					
Pros (at least one)	0.017	0.022	0.577		1.017
Cons (at least one)	−0.152*	0.068	4.945		0.8590
Pros × Cons	0.054	0.102	0.272	1.66(3)	1.055
Verified buyer or not				1.17(1)	
Verified buyer	0.140***	0.015	88.507		1.150
Anonymous reviewer	0	–	–		
Recommendation to a friend				3.90(2)	
Yes	0.069*	0.030	5.355		1.072
No	−0.310***	0.046	46.004		0.733
No recommendation given	0	–	–		
Control variables					
Review age	−0.014	0.008	3.334	1.15(1)	0.986
Average price	−0.453***	0.017	724.422	1.08(1)	0.636
Product category				1.96(6)	
Baby & mom	0.222***	0.043	26.592		1.249
Beauty	−0.300***	0.035	74.010		0.741
Household, food, & pets	0.387***	0.039	99.107		1.472
Medicine & health	0.068	0.036	3.477		1.070
Personal care	0.091*	0.036	6.238		1.095
Sexual well-being	−0.246***	0.048	26.829		0.782
Other	0	–	–		
Seasonality				1.00(4)	
July	−0.136**	0.047	8.470		0.873
August	−0.184***	0.047	15.277		0.832
September	−0.171***	0.047	13.136		0.843
October	−0.103*	0.050	4.329		0.902
June 2014	0	–	–		

Notes. *$p < 0.05$, **$p < 0.01$, ***$p < 0.001$.

that reviews that are perceived as too lengthy hurt purchase probability. Note that the horizontal axis is in log units. The top of the plot shows the original (unlogged) units.

Concerning H2 (i.e. an interaction between review valence and helpfulness), the result shows that there is a positive interaction effect between review valence and review helpfulness on purchase probability, which confirms our hypothesis. Figure 4 shows that the helpfulness slope is steeper for negative reviews than for positive or neutral ones. As a negative review gathers more helpfulness votes, the purchase likelihood decreases, while the effect of helpfulness votes for positive or neutral reviews is flatter. It is surprising that there is a negative association between helpfulness votes and purchase for positive and neutral reviews. This suggests that when individuals are exposed to reviews with positive valence and a higher number of helpfulness votes, they are less likely to make a purchase. Therefore, H2 is only partially confirmed.

Regarding H3 (i.e. message sidedness), we predicted that the presence of pros will increase purchase probability (H3a), whereas the presence of cons will do the opposite and

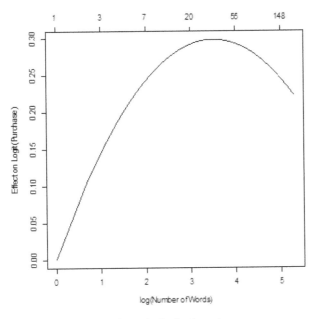

Figure 3. The effect of the number of words on the logit of purchase.

decrease purchase probability (H3b). We did not find any influence of the presence of pros. However, we found a negative impact of the presence of cons. This implies that a heuristic cue signalling negative aspects of a product has a bigger influence than the one with positive aspects. H3 is partially confirmed. Regarding RQ1 (i.e. an interaction between pros and cons), we did not find any significant effect between the presence of pros and cons.

With regard to H4 (i.e. source credibility), we found that displaying a review authored by a verified buyer has a positive influence on purchase probability, supporting previous findings on source expertise and credibility. When consumers are exposed to reviews written by a verified buyer, the odds of product purchase increase by 15%, compared to the case when they read reviews written by anonymous reviewers. H4 is confirmed.

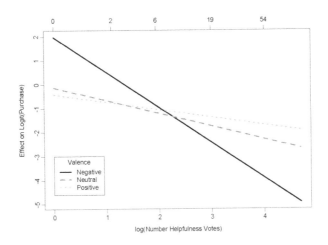

Figure 4. The interaction effect of valence and helpfulness on purchase.

Finally, regarding H5 (i.e. reviewer recommendation), reviewers' willingness to recommend a product to a friend affects purchase probability. We found that compared to when such recommendations are not present, recommendation has a significantly positive impact on purchase probability. Similar to pros and cons, we see a stronger influence of a negative reaction from a reviewer. When consumers see a review that contains 'No' to the question, 'Would you recommend this to a friend?', the odds of purchase decrease by 26.7% compared to when such answers are not provided at all. On the contrary, when a reviewer reports his/her willingness to recommend a product, the odds of purchase increase by 7% compared to no recommendation. Hence, H5 is confirmed.

Discussion

Theoretical Implications

Customers trust online product reviews and hence increasingly consult them to make an informed purchase decision. While the literature on eWOM has been growing quickly, extant studies have focused on the role of review valence, volume, and/or source credibility, but have not devoted much attention to other review features. There has not been much research that has tried to integrate relevant review or reviewer characteristics in a single model. Furthermore, review helpfulness has been overwhelmingly treated as an outcome variable in extant research. There is a dearth of empirical studies that have used individual-level sales data. Most research has focused on proxy measures such as sales rank, attitudes, or purchase intention. To fill these gaps in existing literature, this study applied the HSM and took a comprehensive approach by examining the impact of review and reviewer characteristics as well as possible interaction effects of review features on purchase decisions.

Overall, our findings show that most of the review features we examined significantly influence purchase probability. We found that argument quality (measured as review length) has an inverted-U shaped relationship with purchase probability. This suggests that the positive effect of review length reaches its maximum and diminishes if a review becomes too long. It also confirms recent studies that found a nonlinear effect of review length on outcome variables such as review helpfulness (Baek, Ahn, and Choi 2012; Huang et al. 2015; Kuan et al. 2015). The persuasive effects of message length in an advertising context have shown mixed results. For example, Wells, Leavitt, and McConville (1971) showed that longer commercials presented more product usage illustrations, which increased viewers' opportunity to elaborate on the message. More elaboration leads to more counter-arguing, which ultimately resulted in more negative attitudes toward the advertisement. However, Rethans, Swasy, and Marks (1986) did not find such an effect. There are several explanations for the inverted-U relationship between review length and purchase probability. First, consumers may experience cognitive overload when they perceive a review as lengthy although it is informative. Second, in line with Gossen's diminishing marginal utility law suggesting that the marginal utility of each unit decreases as the supply of unit increases, review readers may predict that a review exceeding a certain length will not provide additional information value, or even perceive it as less informative. Finally, following the ELM and a two-factor perspective (Cacioppo and Petty 1980), longer reviews provide more opportunities for consumers to elaborate on the message and its arguments and enhance counter-arguing, which is detrimental to purchase decisions.

Previous research has pointed out the role of star rating as a heuristic for popularity and quality (Sundar, Oeldorf-Hirsch, and Xu 2008) and its positive impact on outcome variables. What we found in this study is a more complicated picture that shows a moderating impact of review helpfulness. In particular, negative reviews (i.e. with one or two stars) and a large number of helpfulness votes exert a strong negative influence on purchase probability. This may be attributed to the fact that negative reviews are scarce, and thus when negative reviews receive a higher number of helpful votes, their credibility as information sources increases. The effects of positive and neutral reviews are quite flat, which suggests that the level of review helpfulness does not play an important role for positive and neutral reviews. Since the majority of online reviews are positive and consumers expect to see positive reviews (Chen and Lurie 2013), they may process them rather peripherally. When they encounter negative reviews, such reviews are in contrast to the majority of other reviews, which may trigger more system processing of information and lead to a stronger effect of reviews on attitudes and behaviors. In addition, when consumers process information systematically, they look for additional arguments such as star rating or helpfulness votes. Other customers' agreement with the review expressed in helpfulness votes may make the review more credible and hence more persuasive. Following the bias hypothesis of the HSM, negative star rating and many helpfulness votes will then negatively bias central processing (Chaiken and Maheswaran 1994). This process may result in a negative evaluation of the product and hence lower purchase probability.

The results also suggest that the presence of pros does not have a significant impact, but that of cons does decrease purchase probability. Related to the discussion above, this could be again due to the positive bias of online reviews in general (Aral 2014). Since the overall sentiment on the review system is positive, the inclusion of negative information attracts readers' attention (Willemsen et al. 2011) and may be expected to increase reviews' credibility, because negative information may seem more valuable and persuasive (Baumeister et al. 2001). Indeed, some studies have shown that the number of positive online reviews is disproportionately high, which may cause customers to discount positive reviews as not reliable (Chevalier and Mayzlin 2006). In line with this, the accessibility-diagnostic theory predicts that negative information exerts a stronger influence on judgements than positive information (Herr, Kardes, and Kim 1991). Finally, the nonsignificance of the interaction term between pros and cons suggests that two-sided arguments may not be seen persuasive in online product reviews. This is consistent with Schlosser's (2011) argument that one-sided arguments seem more persuasive and helpful particularly for online reviews because the motivation of posting online reviews is to help other consumers by telling them the quality of a product.

The results regarding reviewer characteristics suggest that people pay attention to the nature of review source. Whether a reviewer actually bought and used a product mattered for purchase decision. Extant research shows that people form a more positive attitude toward a product and a website when reviews are written by other consumers than experts or companies (Park, Lee, and Han 2007). Showing a badge that indicates a review is written by a consumer who has purchased and used the product increases purchase probability. In addition, a statement whether a reviewer is willing to recommend a product to his/her friend increases purchase probability compared to when there is no intention of recommendation. On the contrary, expressing unwillingness to recommend a product has a negative influence compared to not expressing any intention to recommend the product. The intention to share positive word-of-mouth (WOM) about a product is a significant indicator of brand loyalty (Dick and Basu 1994). Thus, the cue showing that a reviewer is willing to recommend a product can be crucial information that signals the quality of the product.

Practical implications

From a managerial perspective, the findings of this study suggest that companies should consider which aspects of reviews must be taken more seriously. This also has an important implication for the design of the websites and reviewing requests sent to customers. First, longer reviews generally provide more information, but they may want to limit the length of each review by finding an optimal review length. Website designers may want to prime customers with the average number of words other customers wrote, for example. Second, when companies notice clearly negative heuristics, for instance, negative star ratings, a list of cons, or a 'No' to the question whether they are willing to recommend a product to a friend, they should find a way to intervene and identify where the cause of dissatisfaction lies. Hence, companies should consider engaging in conversations with customers through webcare. Third, companies should actively solicit reviewers from customers who actually purchased and used a product (Askalidis, Kim, and Malthouse 2017). The problem of verified purchase is that it only recognizes consumers who bought an item from a particular retailer that provides the online review system. For consumers who made a purchase elsewhere, there is no way that they can visually show that they are verified users. Thus, giving an option to reviewers to show that they have a real experience of using a product by for example indicating where they got the product from is a good way to potentially increase the overall trustworthiness of reviews.

The aforementioned point about source credibility provides further insights to advertising scholars and practitioners. WOM has been considered as a key to advertising effectiveness (Keller and Fay 2012), so is eWOM in the era of Web 2.0. When it comes to eWOM, customers often have difficulty evaluating credibility of messages (Levy and Gvili 2015). There have been debates in the advertising discipline regarding whether brands should encourage consumers to post a product review, and, if so, which type of platform would be the most effective. Advertisers sometimes compensate consumers for posting a product review on their personal blogs believing that the review is perceived more reliable. Previous research shows that brand- and consumer-generated websites are considered equally persuasive when consumers read positive reviews (Xue and Phelps 2004). Alternatively, a study by Lee and Youn (2009) finds that customers are more likely to recommend a product after reading a positive review on a brand-generated website than consumer-generated one (i.e. a blog), suggesting that consumers are suspicious about reviews in consumer-generated websites. Perceived source trustworthiness positively affects eWOM's influence on decision-making (Lopez and Sicilia 2014). Hence, it is of paramount importance that consumers know that the reviewer is expressing his/her honest opinion. This is also supported by our results showing that information whether a reviewer actually bought and used a product is important for customers' purchase decision. In sum, we think that advertisers should stimulate customers' participation in eWOM while providing transparent information about the reviewing process to diminish consumers' suspicion.

Limitations and suggestions for future research

Although this study provides insights regarding how each characteristic of online product reviews influences purchase probability, a few limitations should be noted and addressed in future research. First, as we mentioned in the Method section, we had to limit our sample to products with one review due to the way the data were collected. This reduces our ability to generalize the findings, but allows us to link consumers' exposure to reviews with purchase behavior. Second, argument quality was measured as review length, a proxy variable. Without a content analysis of argument strength in each review, this study was not able to test the effect of the level of argument quality on purchase decision. Third, this study used data from a single online retailer that mostly sells search goods and consumer packaged goods. Consumers' interest in reviews and their persuasive power may depend on product categories (Allsop, Bassett, and Hoskins 2007). Finally, due to the nature of a secondary analysis of existing data, we did not have all the variables we hoped to include in the regression model such as personal characteristics, consumption history, WOM activities, or brands' marketing efforts.

Future research should investigate other content and source characteristics that are not observed in this study. In particular, in-depth analyses of review content – its readability, relevance, or informativeness – can provide further insights regarding the financial impact of review content (Mackiewicz and Yeats 2014; Park and Nicolau 2015). To better understand how consumers process online reviews, future studies should include content analysis and categorize different types of arguments presented in reviews. Also, many studies investigating online reviews (including our study) build on dual-process theories to establish their theoretical framework. Future research needs to go one step further and actually test dual-process theories in the eWOM context (e.g. Zhang et al. 2014). This would require a controlled experimental design. In addition, the role of aggregated information of reviews for a single product (i.e. a summary section) should also be investigated. It is possible that consumers may not read individual reviews, but read a summary section presented at the top of the Review tab. Whether a summary section exerts a bigger influence than individual reviews or whether there is a condition under which a summary of all reviews and individual reviews have an interaction (for instance, high consistency) is an interesting question that remains to be investigated. With regard to reviewer characteristics, understanding how the similarity between reviewers and review readers influences purchase decision can provide interesting insights.

In sum, this study contributes to the literature on online product reviews and consumer-generated advertising by looking at the effect of different review features on purchase decisions. It confirms the effects of review characteristics that are identified based on the HSM approach (e.g. Cheung and Thadani 2012). Some findings may seem intuitive, for instance, the positive effect of a verified buyer badge or the presence of cons. We also have unexpected findings, for example, the moderating effect of review helpfulness or the curvilinear effect of review length, which requires further scholarly investigation.

Follow-up experiments that corroborate the findings from this research and studies that examine the effects of review features on purchase decisions in other product or service categories would help generalize the findings on different effects of various review features.

Notes

1. We assessed multicollinearity using GVIFs proposed by Fox and Monette (1992). GVIFs are an appropriate way to assess multicollinearity in models with categorical predictors and polynomials. When there is multicollinearity, slope estimates are unbiased but have inflated standard errors (i.e. the variance of the estimates is inflated), but standard errors are also determined by sample size, which is very large in our case.
2. The model does not overfit the data. Our logistic regression model is estimated on 413,666 observations and has 25 parameters. Perhaps, the greatest risk to overfitting comes from the dummy variables, but we have large samples (valence has 63,979, 28,253, and 322,434 values; review recommendation has 29,274, 64,170 and 320,222 values; category has 18,542, 145,620, 29,515, 112,842, 17,670, 62,341, and 27,136; seasonality has 8224, 142,476, 111,599, 108,494, and 43,873 observations). To make sure that we are not overfitting, we use five-fold cross validation by assigning a random value 1–5 to each case (Kuhn and Johnson 2013, 69–70). We then estimated our model five times, each time leaving out one part, and applied the model to the left-out part. The value of AUC was computed on the held-out values. The original AUC value was 0.6146. The five-fold CV value is 0.6138, which differs by 0.0008. We also tried 10-fold cross validation and got the same AUC value to four decimal places. Thus, there is no evidence for overfitting.
3. For a predictor having a single degree of freedom, GVIF equals VIF (equaling $1/(1 - R^2)$ from a model predicting X_j from the other predictors in the model). Let X_1 be a block of predictors (e.g. multiple dummies for a categorical predictor), X_2 be a block of the remaining predictors in the model, and X be all predictors (bind X_1 and X_2 to product X). Let R_1 be the correlation matrix of X_1, R_2 the correlation matrix for X_2 and R be the correlation matrix for X. Then, GVIF = det $(R_1)^*\det(R_2)/\det(R)$. A benefit of GVIF is that it is invariant to the choice of the baseline value (e.g. negative for valence), or category (requiring six dummies).

Acknowledgments

The authors thank the IMC Medill Spiegel Digital & Database Research Center for granting access to the data for this study.

Disclosure statement

No potential conflict of interest was reported by the authors.

ORCID

Su Jung Kim ⓘ http://orcid.org/0000-0003-2025-4019

References

Allsop, D.T., B.R. Bassett, and J.A. Hoskins. 2007. Word-of-mouth research: Principles and applications. *Journal of Advertising Research* 47, no. 4: 398–411. doi:10.2501/s0021849907070419

Aral, S. 2014. The problem with online ratings. *MIT Sloan Management Review* 55, no. 2: 47–52.

Askalidis, G., S.J. Kim, and E.C. Malthouse. 2017. Understanding and overcoming biases in online review systems. *Decision Support Systems* 97: 23–30. doi:10.1016/j.dss.2017.03.002

Baber, A., R. Thurasamy, M.I. Malik, B. Sadiq, S. Islam, and M. Sajjad 2016. Online word-of-mouth antecedents, attitude and intention-to-purchase electronic products in Pakistan. *Telematics and Informatics* 33, no. 2: 388–400.

Baek, H., J. Ahn, and Y. Choi. 2012. Helpfulness of online consumer reviews: Readers' objectives and review cues. *International Journal of Electronic Commerce* 17, no. 2: 99–126. doi:10.2753/JEC1086-4415170204

Baumeister, R.F., E. Bratslavsky, C. Finkenauer, and K.D. Vohs. 2001. Bad is stronger than good. *Review of General Psychology* 5, no. 4: 323–70. doi:10.1037/1089-2680.5.4.323

Blazevic, V., W. Hammedi, I. Garnefeld, R.T. Rust, T. Keiningham, T.W. Andreassen, N. Donthu, and W. Carl. 2013. Beyond traditional word−of−mouth: An expanded model of customer−driven influence. *Journal of Service Management* 24, no. 3: 294–313. doi:10.1108/09564231311327003

Bohner, G., S. Chaiken, and P. Hunyadi. 1994. The role of mood and message ambiguity in the interplay of heuristic and systematic processing. *European Journal of Social Psychology* 24, no. 1: 207–21.

Bosman, D.J., C. Boshoff, and G.-J. van Rooyen. 2013. The review credibility of electronic word-of-mouth communication on e-commerce platforms. *Management Dynamics* 22, no. 3: 29–44.

Cacioppo, J.T., and R.E. Petty. 1980. Persuasiveness of communications is affected by exposure frequency and message quality: A theoretical and empirical analysis of persisting attitude change. *Current Issues & Research in Advertising* 3, no. 1: 97–122.

Chaiken, S. 1980. Heuristic versus systematic information processing and the use of source versus message cues in persuasion. *Journal of Personality and Social Psychology* 39, no. 5: 752–66. doi:10.1037/0022-3514.39.5.752

Chaiken, S., and D. Maheswaran. 1994. Heuristic processing can bias systematic processing: Effects of source credibility, argument ambiguity, and task importance on attitude judgment. *Journal of Personality and Social Psychology* 66, no. 3: 460–473. doi:10.1037/0022-3514.66.3.460

Chen, Y.-F. 2008. Herd behavior in purchasing books online. *Computers in Human Behavior* 24, no. 5: 1977–92. doi:10.1016/j.chb.2007.08.004

Chen, Z., and N.H. Lurie. 2013. Temporal contiguity and negativity bias in the impact of online word of mouth. *Journal of Marketing Research* 50, no. 4: 463–76. doi:10.1509/jmr.12.0063

Cheng, Y.-H., and H.-Y. Ho. 2015. Social influence's impact on reader perceptions of online reviews. *Journal of Business Research* 68, no. 4: 883–7. doi:10.1016/j.jbusres.2014.11.046

Cheung, C.M.K., M.K.O. Lee and N. Rabjohn. 2008. The impact of electronic word-of-mouth. *Internet Research* 18, no. 3: 229–47.

Cheung, C.M.-Y., C.-L. Sia, and K.K.Y. Kuan. 2012. Is this review believable? A study of factors affecting the credibility of online consumer reviews from an ELM perspective. *Journal of the Association for Information Systems* 13, no. 8: 618–35.

Cheung, C.M.K., and D.R. Thadani. 2012. The impact of electronic word-of-mouth communication: A literature analysis and integrative model. *Decision Support Systems* 54, no. 1: 461–70. doi:10.1016/j.dss.2012.06.008

Chevalier, J.A., and D. Mayzlin. 2006. The effect of word of mouth on sales: Online book reviews. *Journal of Marketing Research* 43, no. 3: 345–54. doi:10.1509/jmkr.43.3.345

Chintagunta, P.K., S. Gopinath, and S. Venkataraman. 2010. The effects of online user reviews on movie box office performance: Accounting for sequential rollout and aggregation across local markets. *Marketing Science* 29, no. 5: 944–57. doi:10.1287/mksc.1100.0572

Clemons, E.K., G.G. Gao, and L.M. Hitt. 2006. When online reviews meet hyperdifferentiation: A study of the craft beer industry. *Journal of Management Information Systems* 23, no. 2: 149–71.

Dhanasobhon, S., P.-Y. Chen, M. Smith, and P.-y Chen. 2007. An analysis of the differential impact of reviews and reviewers at Amazon.com. Paper presented at the ICIS 2007 proceedings in Montreal, Canada.

Dick, A.S., and K. Basu. 1994. Customer loyalty: Toward an integrated conceptual framework. *Journal of the Academy of Marketing Science* 22, no. 2: 99–113. doi:10.1177/0092070394222001

Doh, S.-J., and J.-S. Hwang. 2009. How consumers evaluate eWOM (electronic word-of-mouth) messages. *CyberPsychology & Behavior* 12, no. 2: 193–7. doi:10.1089/cpb.2008.0109

Dou, X., J.A. Walden, S. Lee, and J.Y. Lee. 2012. Does source matter? Examining source effects in online product reviews. *Computers in Human Behavior* 28, no. 5: 1555–63. doi:10.1016/j.chb.2012.03.015

Filieri, R. and F. Mcleay. 2013. E-wom and accommodation: An analysis of the factors that influence travelers' adoption of information from online reviews. *Journal of Travel Research* 53, no. 1: 44–57.

Finn, A., L. Wang, and T. Frank. 2009. Attribute perceptions, customer satisfaction and intention to recommend e-services. *Journal of Interactive Marketing* 23, no. 3: 209–20. doi:10.1016/j.intmar.2009.04.006

Floyd, K., R. Freling, S. Alhoqail, H.Y. Cho, and T. Freling. 2014. How online product reviews affect retail sales: A meta-analysis. *Journal of Retailing* 90, no. 2: 217–32. doi:10.1016/j.jretai.2014.04.004

Fox, J., and G. Monette. 1992. Generalized collinearity diagnostics. *Journal of the American Statistical Association* 87, no. 417: 178–83. doi:10.1080/01621459.1992.10475190

Gopinath, S., J.S. Thomas, and L. Krishnamurthi. 2014. Investigating the relationship between the content of online word of mouth, advertising, and brand performance. *Marketing Science* 33, no. 2: 241–58. doi:10.1287/mksc.2013.0820

Gupta, P. and J. Harris. 2010. How e-wom recommendations influence product consideration and quality of choice: A motivation to process information perspective. *Journal of Business Research* 63, no. 9–10: 1041–49.

Herr, P.M., F.R. Kardes, and J. Kim. 1991. Effects of word-of-mouth and product-attribute information on persuasion: An accessibility-diagnosticity perspective. *Journal of Consumer Research* 17, no. 4: 454–62. doi:10.1086/208570

Huang, A.H., K. Chen, D.C. Yen, and T.P. Tran. 2015. A study of factors that contribute to online review helpfulness. *Computers in Human Behavior* 48: 17–27. doi:10.1016/j.chb.2015.01.010

Kahneman, D. 1973. *Attention and effort*. Englewood Cliffs, NJ: Prentice-Hall.

Keller, E., and B. Fay. 2012. Word-of-mouth advocacy: A new key to advertising effectiveness. *Journal of Advertising Research* 52, no. 4: 459–64.

Kim, H.-S., P. Brubaker, and K. Seo. 2015. Examining psychological effects of source cues and social plugins on a product review website. *Computers in Human Behavior* 49: 74–85. doi:10.1016/j.chb.2015.02.058

King, R.A., P. Racherla, and V.D. Bush. 2014. What we know and don't know about online word-of-mouth: A review and synthesis of the literature. *Journal of Interactive Marketing* 28, no. 3: 167–83. doi:10.1016/j.intmar.2014.02.001

Kuhn, M. and K. Johnson. 2013. Ed. Johnson, K, Springerlink and Link. *Applied predictive modeling.* New York, NY: Springer.

Kuan, K.K.Y., H. Kai-Lung, P. Prasarnphanich, and L. Hok-Yin. 2015. What makes a review voted? An empirical investigation of review voting in online review systems. *Journal of the Association for Information Systems* 16, no. 1: 48–71.

Lee, M., and S. Youn. 2009. Electronic word of mouth (eWOM). *International Journal of Advertising* 28, no. 3: 473–99. doi:10.2501/S0265048709200709

Levy, S., and Y. Gvili. 2015. How credible is e-word of mouth across digital-marketing channels? The roles of social capital, information richness, and interactivity. *Journal of Advertising Research* 55, no. 1: 95–109. doi:10.2501/jar-55-1-095-109

Li, M., L. Huang, C.-H. Tan, and K.-K. Wei. 2013. Helpfulness of online product reviews as seen by consumers: Source and content features. *International Journal of Electronic Commerce* 17, no. 4: 101–36. doi:10.2753/jec1086-4415170404

Li, X., and L.M. Hitt. 2008. Self-selection and information role of online product reviews. *Information Systems Research* 19, no. 4: 456–74. doi:10.1287/isre.1070.0154

Liu-Thompkins, Y., and E.C. Malthouse. 2017. A primer on using behavioral data for testing theories in advertising research. *Journal of Advertising* 46, no. 1: 213–25. doi:10.1080/00913367.2016.1252289

Liu, Y. 2006. Word of mouth for movies: Its dynamics and impact on box office revenue. *Journal of Marketing* 70, no. 3: 74–89. doi:10.2307/30162102

López, M. and M. Sicilia. 2014. Ewom as source of influence: The impact of participation in ewom and perceived source trustworthiness on decision making. *Journal of Interactive Advertising* 14, no. 2: 86–97.

Mackiewicz, J., and D. Yeats. 2014. Product review users' perceptions of review quality: The role of credibility, informativeness, and readability. *IEEE Transactions on Professional Communication* 57, no. 4: 309–24. doi:10.1109/TPC.2014.2373891

Maslowska, E., E.C. Malthouse, and S.F. Bernritter. 2017. Too good to be true: The role of online reviews' features in probability to buy. *International Journal of Advertising* 36, no. 1: 142–63. doi:10.1080/02650487.2016.1195622

Maslowska, E., E.C. Malthouse, and V. Viswanathan. 2017. Do customer reviews drive purchase decisions? The moderating roles of review exposure and price. *Decision Support Systems* 98: 1–9. doi:10.1016/j.dss.2017.03.010

McCroskey, J.C., V.P. Richmond, and J.A. Daly. 1975. The development of a measure of perceived homophily in interpersonal communication. *Human Communication Research* 1, no. 4: 323–32. doi:10.1111/j.1468-2958.1975.tb00281.x

Metzger, M.J., A.J. Flanagin, and R.B. Medders. 2010. Social and heuristic approaches to credibility evaluation online. *Journal of Communication* 60, no. 3: 413–39. doi:10.1111/j.1460-2466.2010.01488.x

Mudambi, S.M., and D. Schuff. 2010. What makes a helpful review? A study of customer reviews on Amazon.com. *MIS Quarterly* 34, no. 1: 185–200.

Otterbacher, J. 2009. 'Helpfulness' in online communities: A measure of message quality. Paper presented at the proceedings of the SIGCHI conference on human factors in computing systems, Boston, MA, USA.

Park, D.-H., and S. Kim. 2008. The effects of consumer knowledge on message processing of electronic word-of-mouth via online consumer reviews. *Electronic Commerce Research and Applications* 7, no. 4: 399–410. doi:10.1016/j.elerap.2007.12.001

Park, D.-H., J. Lee, and I. Han. 2007. The effect of on-line consumer reviews on consumer purchasing intention: The moderating role of involvement. *International Journal of Electronic Commerce* 11, no. 4: 125–48. doi:10.2753/JEC1086-4415110405

Park, S., and J.L. Nicolau. 2015. Asymmetric effects of online consumer reviews. *Annals of Tourism Research* 50: 67–83. doi:10.1016/j.annals.2014.10.007

Petty, R.E., and J.T. Cacioppo. 1986. *Communication and persuasio: Central and peripheral routes to attitude change*. New York: Springer-Verlag.

Purnawirawan, N., M. Eisend, P. De Pelsmacker, and N. Dens. 2015. A meta-analytic investigation of the role of valence in online reviews. *Journal of Interactive Marketing* 31: 17–27. doi:10.1016/j.intmar.2015.05.001

Quaschning, S., M. Pandelaere, and I. Vermeir. 2015. When consistency matters: The effect of valence consistency on review helpfulness. *Journal of Computer-Mediated Communication* 20, no. 2: 136–52. doi:10.1111/jcc4.12106

Reichheld, F.F. 2003. The one number you need to grow (cover story). *Harvard Business Review* 81, no. 12: 46–54.

Rethans, A.J., J.L. Swasy, and L.J. Marks. 1986. Effects of television commercial repetition, receiver knowledge, and commercial length: A test of the two-factor model. *Journal of Marketing Research* 23, no. 1: 50–61. doi:10.2307/3151776

Riegner, C. 2007. Word of mouth on the Web: The impact of Web 2.0 on consumer purchase decisions. *Journal of Advertising Research* 47, no. 4: 436–47.

Rucker, D.D., Z.L. Tormala, R.E. Petty, and P. Briñol. 2014. Consumer conviction and commitment: An appraisal-based framework for attitude certainty. *Journal of Consumer Psychology* 24, no. 1: 119–36. doi:10.1016/j.jcps.2013.07.001

Schindler, R.M., and B. Bickart. 2012. Perceived helpfulness of online consumer reviews: The role of message content and style. *Journal of Consumer Behaviour* 11, no. 3: 234–43. doi:10.1002/cb.1372

Schlosser, A.E. 2011. Can including pros and cons increase the helpfulness and persuasiveness of online reviews? The interactive effects of ratings and arguments. *Journal of Consumer Psychology* 21, no. 3: 226–39. doi:10.1016/j.jcps.2011.04.002

Sher, P.J. and L. Sheng-Hsien. 2009. Consumer skepticism and online reviews: An elaboration likelihood model perspective. *Social Behavior & Personality* 37, no. 1: 137–43.

Sridhar, S., and R. Srinivasan. 2012. Social influence effects in online product ratings. *Journal of Marketing* 76, no. 5: 70–88. doi:10.1509/jm.10.0377

Sundar, S.S. 2008. The MAIN model: A heuristic approach to understanding technology effects on credibility. In *Digital media, youth, and credibility*, eds. M.J. Metzger and A.J. Flanagin, 73–100. Cambridge, MA: MIT Press.

Sundar, S.S., A. Oeldorf-Hirsch, and Q. Xu. 2008. The bandwagon effect of collaborative filtering technology. Paper presented at the CHI'08 extended abstracts on human factors in computing systems in Florence, Italy.

The Nielsen Company. 2015. Recommendations from friends remain most credible form of advertising among consumers; branded websites are the second-highest-rated form. <http://www.nielsen.com/us/en/press-room/2015/recommendations-from-friends-remain-most-credible-form-of-advertising.html>

Todorov, A., S. Chaiken, and M.D. Henderson. 2002. The heuristic-systematic model of social information processing. In *The persuasion handbook: Developments in theory and practice*, ed. J.P. Dillard and M. Pfau, 195–212. Thousand Oaks, CA: SAGE.

Vasa, R., L. Hoon, K. Mouzakis, and A. Noguchi. 2012. A preliminary analysis of mobile app user reviews. Paper presented at the proceedings of the 24th Australian computer–human interaction conference, Melbourne, Australia.

Walther, J.B., Y. Liang, T. Ganster, D.Y. Wohn, and J. Emington. 2012. Online reviews, helpfulness ratings, and consumer attitudes: An extension of congruity theory to multiple sources in web 2.0. *Journal of Computer-Mediated Communication* 18, no. 1: 97–112. doi:10.1111/j.1083-6101.2012.01595.x

Wells, W.D., C. Leavitt, and M. McConville. 1971. A reaction profile for TV commercials. *Journal of Advertising Research* 11, no. 6: 11–18.

Willemsen, L.M., P.C. Neijens, F. Bronner, and J.A. de Ridder. 2011. "Highly recommended!" The content characteristics and perceived usefulness of online consumer reviews. *Journal of Computer-Mediated Communication* 17, no. 1: 19–38. doi:10.1111/j.1083-6101.2011.01551.x

Xue, F., and J.E. Phelps. 2004. Internet-facilitated consumer-to-consumer communication: the moderating role of receiver characteristics. *International Journal of Internet Marketing and Advertising* 1, no. 2: 121–36. doi:10.1504/ijima.2004.004016

Zhang, K.Z.K., S.J. Zhao, C.M.K. Cheung, and M.K.O. Lee. 2014. Examining the influence of online reviews on consumers' decision-making: A heuristic–systematic model. *Decision Support Systems* 67: 78–89. doi:10.1016/j.dss.2014.08.005

Zhang, W., and S.A. Watts. 2008. Capitalizing on content: Information adoption in two online communities. *Journal of the Association for Information Systems* 9, no. 2: 72–93.

Zhang, X., and C. Dellarocas. 2006. The lord of the ratings: Is a movie's fate is influenced by reviews? In *ICIS 2006 proceedings*, 117. Milwaukee, WI: ICIS.

Preannouncement messages: impetus for electronic word-of-mouth

Hao Zhang and Yung Kyun Choi

ABSTRACT

To enhance public opinion, firms must establish effective communication strategies before introducing new products to the market. One of the most popular strategies is to make new product preannouncements (NPPs) to attract consumer attention and create positive buzz. The authors of this paper offer theoretical insights into the relationship between NPPs and electronic word-of-mouth (eWOM). They conduct three experiments to investigate the effects of NPP message clarity and brand characteristics on eWOM. In Study 1, they find that consumers are more likely to disseminate eWOM when NPPs have high rather than low clarity. In Study 2, they show that brand familiarity moderates the effects of NPP message clarity. In Study 3, they show that brand preference moderates the effects. Marketers can use the findings to establish more effective NPP communication strategies that signal brand information. Theoretical implications are discussed.

Introduction

Information technology enables modern consumers to obtain information about new products even before they are launched. For example, before Apple launched the new iPhone 7, the Internet had already provided much information about its attributes and appearance. Consequently, consumers had already formed their first perceptions through a vast array of electronic word-of-mouth (eWOM) that may have had huge impacts on consumers' decision process. However, such information about new products may be incomplete, or even faked, so that opinion leaders can easily form negative first impressions (Haenlein and Libai 2013). Thus, to create positive public opinion, firms must establish effective communication strategies before introducing their new products to the market. One of their best choices is to make new product preannouncements (NPPs) to create positive buzz.

NPPs are important prelaunch marketing communications techniques (Chen and Wong 2012) and efficient strategies to gain competitive advantages, especially in markets characterized by intense innovative and competition. NPPs are one of the most popular tools for signalling product quality, creating early customer awareness, encouraging potential purchases, and shortening product lifecycles (Eliashberg and Robertson 1988; Lilly and Walters 1997; Mishra and Bhabra 2001; Su and Rao 2010).

The clarity of advertising messages is a vital factor in communication success (Kim, Han, and Yoon 2010; Taylor, Franke, and Bang 2006). In this regard, NPP specificity is a strategic signal directed at customers, and thus positively influences shareholder value (Sorescu, Shankar, and Hushwaha 2007). Also, signalling is most useful when consumers know little about product quality before purchase (Kirmani and Rao 2000). Thus, NPP message types and brand characteristics (e.g. well known vs. unknown brands) may determine effects on eWOM. Prior studies of NPPs have focused on market influences (e.g. Eliashberg and Robertson 1988), benefits or risks (e.g. Ofek and Turut 2013; Sorescu, Shanker, and Hushwaha 2007), proper timing (e.g. Lilly and Walters 1997), and effects of message features on eWOM (e.g. Yap, Soetarto, and Sweeney 2013).

Although message types and brand characteristics are expected to play interchangeable key roles in successful NPP campaigns, few researchers have explored their simultaneous roles. Therefore, our purpose in this research is to investigate how NPP message clarity and brand characteristics (e.g. brand familiarity and brand preference) interplay to influence eWOM across NPP communication contexts. We expect to help scholars and marketers understand how NPP message format and brand characteristics influence active communications regarding promotional messages through eWOM. We also provide insight into signalling theory by identifying potential moderators for NPP communication and their consequences on eWOM. Finally, we expect our results to have marketing and managerial implications for making NPP campaigns more effective.

Theoretical background and hypotheses

New product preannouncement as a signalling tool

Preannouncement, a form of market signalling, is defined as 'formal, deliberate communication before a firm actually undertakes a particular marketing action such as a price change, a new advertising campaign, or a product line change' (Eliasherberg and Robertson 1988, 282). Especially in competitive high-tech industries, firms prefer to use NPPs to signal new strategies in pricing, distribution channels, or advertising (Su and Rao, 2010).

NPPs can be easily confused with new product announcements (NPAs). Basically, NPAs are announcements made within one month before launch; NPPs are made more than one month before launch (Su and Rao 2010). However, this simplistic typology needs expansion. NPPs and NPAs are not isolated market-signalling activities: they are both dynamic components of a communication process. As Figure 1 shows, NPPs are not one-time information releases; they are continually officially updated, precede NPAs, and include some uncertainties. NPPs usually omit details or final features of new products until they are actually available. But as the time for NPAs grows closer, NPPs provide more details about attributes, launch dates, or even price ranges for the new products. Later, the firm may release additional information and interact with customers through social network sites, professional evaluations, or lead-user opinions.

Signalling theory can be used to explain organizational procedures for conveying product information, intentions, and abilities (Garcia-Granero et al. 2015). Marketing communication signals tell buyers about company characteristics or products; buyers then examine the communications to evaluate whether the qualities are credible and

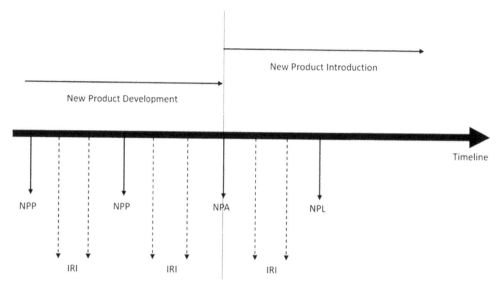

NPP: New Product Preannouncement
NPA: New Product Announcement
NPL: New Product Launch
IRI: Information Release Incidents

Figure 1. New product launch process.

valid (Mavlanova, Benhunan-Fich, and Koufaris 2012). If companies choose to falsify infor-mation, they must feel that the benefits will surpass the possible costs of deception (Donath 2007).

NPPs can secure distribution channels: if distributors or retailers are pre-informed about new products, they can be better prepared for distribution, promotion, and preparation of a supportive environment (Wind and Mahajan 1987). Firms can use NPPs to alert the stock market and shareholders, which then often yields long-term firm benefits (Sorescu, Shankar, and Kushwaha 2007), but the relationship is indirect and depends on how rapidly and efficiently competitors can revise their strategies to counter the preannouncements (Su and Rao 2010). In intensive competition, NPPs can signal customers to delay purchas-ing from competitors and wait instead for more desirable upcoming products (Bayus, Jain, and Rao 2001).

Although NPP effects on different target audiences have been researched, firms estab-lishing NPP strategies must focus on customers as the key factors and must set customer-specific objectives (Chen and Wong 2012). NPPs can evoke curiosity and interest to accel-erate the initial rate of customer adoption (Lilly and Krishnan 1996), reduce adoption resis-tance (Gatignon and Robertson 1991), lower switching costs (Eliashberg and Robertson 1988), and enhance brand image (Rabino and Moore 1989). Besides inspiring customer affection, NPPs can generate favourable WOM among opinion leaders, customer innova-tors, and market mavens (Goldsmith et al. 2003).

Electronic word-of-mouth

Word-of-mouth (WOM) refers to the flow of communication about products and services among consumers (De Angelis et al. 2012; Westbrook 1987) and is well known to influence consumer decisions (Chu and Kim 2011; Gilly et al. 1998). Consumers usually consider WOM to be independent, trustworthy, and more credible than messages sent by marketers (Lau and Ng 2001).

The emergence of Internet-based technology has facilitated the development of online electronic word-of-mouth (eWOM) (Chu and Kim 2011), in which potential, actual, or former customers disseminate positive or negative statements about products or companies to multitudes of people and institutions via the Internet (Hennig-Thurau et al. 2004). E-WOM can be generated through a wide range of online channels, such as blogs, emails, twitters, forums, and brand communities (de Angelis et al, 2012). E-WOM is a uniquely independent, anonymous, convenient, and safe way for consumers to express their opinions (Goldsmith and Horowitz 2006; Lee and Youn, 2009; Sen and Lerman, 2007). Thus, eWOM platforms are a popular way for consumers to disseminate information or learn about other consumers' experiences (Duhan et al. 1997).

Several factors motivate consumers to engage in eWOM, such as self-enhancement, innovativeness, opinion leadership, ability, and self-efficacy (King et al. 2014). Although message format is known to influence eWOM (Choi, Seo, and Yoon 2017), most research has limited the scope to motivational factors (Shin et al. 2014) and has rarely focused on sources that will generate eWOM.

Hypothesis development

NPPs and message clarity

The prime concern in all marketing communications is clarity (Fill 1995). Some firms provide detailed preannouncements; others deliver limited information (Lilly and Walters 2000). However, few empirical studies have considered NPP 'message' aspects (Chen and Wong 2012), particularly the degree of message clarity as an essential component in NPP campaigns.

Message clarity indicates whether messages are communicated without ambiguity or noise (Daft et al. 1987; Gatignon and Robertson 1991). Clear messages reduce uncertainty and are more persuasive (Gatignon and Robertson 1991). Indeed, companies must assure 'preannouncement specificity' if they are to reduce investor uncertainty associated with forthcoming products and to generate positive financial returns (Sorescu, Shankar, and Kushwaha 2007). That is, to enhance the credibility of brand signal and decrease perceived risk to the brand, brand signals must provide clear and specific details (Erdem and Swait 1998).

Clear signals have been shown to positively affect target audience responses (Koku 1997); deepen their understandings about new products (Heil and Roberson 1991; Robertson and Rymon 1995), increase their intentions to learn about and purchase products (Kohli 1999), and increase their awareness, interest, and trust of the new products (Heil and Walters 1993; Koku 1997). Consumers may feel more confident about and more likely to disseminate NPP information that conveys highly clear brand signals. Thus:

H1: EWOM will be more likely for NPPs that have high rather than low message clarity.

NPPs and brand characteristics

In traditional offline environments, consumers can access information about product quality through trial or direct product experiences. In contrast, online environments limit the ability to convey intrinsic product attributes such as touch or smell (Grewal et al. 2004). Because online marketing channels offer asymmetric information, they may be more difficult venues for signalling the quality of unfamiliar brands to drive positive eWOM.

Brand familiarity indicates that consumers have accumulated brand-related experiences (Alba and Hutchinson 1987) that have aroused positive and unique consumer associations (Aaker 2002). Highly familiar brands have acquired brand knowledge and brand associations in the general consumer memory (Campbell and Keller 2003). Consequently, consumers have stored knowledge about familiar brands.

Consequently, brand familiarity may influence consumer processing and evaluations (e.g. Hoyer and Brown 1990; Sundaram and Webster 1999). Shoppers tend to spend less time deciding whether to purchase familiar brands; shopping for unfamiliar brands requires more effort (Biswas 1992). Thus, consumers may find it easier to process NPP messages regarding familiar brands.

NPP message clarity may be less essential for familiar brands because consumers already know a great deal about well-known brands. Therefore, signalling and NPP message clarity may be most essential for brands of unknown quality and for products about which consumers are relatively uninformed (Kirmani and Rao 2000). A study of advertising through short messaging service (SMS), message clarity was found to have positive effects on consumer attitudes and acceptance of SMS advertising but negative effects on brand familiarity (Khasawneh and Shuhaiber 2013). The results indicated that SMS brand exposure is more beneficial for unfamiliar brands than for familiar brands. Similarly, a study of in-game advertising showed that after game-players finished playing the game, they had more favourable attitudes toward unfamiliar rather than familiar brands (Mau, Silberer, and Constien 2008).

Consequently, detailed NPPs that provide information about unfamiliar brands will be more appealing for consumers who want to reduce perceived risk or uncertainty before making purchases. Thus:

H2: Brand familiarity will moderate the influence of NPP message clarity on eWOM: high message clarity will be more effective for unfamiliar brands than for familiar brands.

Brand preference, another influencing factor for effective NPP messages, means that consumers have high psychological cognition regarding a brand, evaluate it highly in comparison with other brands, and are highly willing to share information about their preferred brands with friends (Chen and Chang 2008) through eWOM. Brand preference reflects cognitive judgments and affective feelings such as infused value or credibility (Erdem and Swait 1998; Jamal and Al-Marri 2007). Subjective brand preference alters the likelihood of perceptual processing and affect toward brand information. For example, consumers have been shown to have conscious perceptions regarding highly preferred brands (Ramsøy and Skov 2014), but brand preference can be a gradually cumulative process. First, customers evaluate products by value factors such as service and brand image. After they purchase and use products, they decide whether they like the products and want to buy them again. If indeed they form brand preference, they will be more trusting of brand information and more likely to spread positive eWOM (Choe and Zhao 2013;

Jamal and Goode 2001). They will be comparatively less likely to share information about non-preferred brands because they perceive higher risk or brand uncertainty. NPP message clarity then has an important role in increasing brand trust or intentions to purchase non-preferred brands. However, if consumers have already built value and trust for their preferred brands, NPP message clarity can have limited effect. Consequently, highly clear messages conveying brand credibility for non-preferred brands that carry high perceptions of risk and uncertainty may be most beneficial. Thus:

H3: Brand preference will moderate the influence of NPP message clarity on eWOM: high message clarity will be more effective for non-preferred brands than for preferred brands.

Overview of studies

We conducted three experiments to investigate NPP effects on eWOM. We presented participants with product decision scenarios that included information about product preannouncements, provided different types of NPP messages, and used different brand names. In Study 1, we examined the effects of NPP message clarity and found that high clarity had more positive effects on eWOM, while low clarity was less effective. In Study 2, we investigated whether brand familiarity moderates the effects of clarity on eWOM. In Study 3, we explored whether brand preference moderates the effects of NPP message clarity on eWOM.

Pilot study – the effect of NPP message

Although NPP has been shown to positively influence consumer decisions (e.g. Su and Rao 2011), empirical research has been absent. Product diffusion theory indicates that NPP is the first step for launching products. Therefore, before we conducted the main experiments, we did a pilot test to examine whether a NPP message (official preannouncement) is actually more efficient than online news (e.g. news announcement with leaked pictures of new products).

Method

Participants were 60 undergraduate business school juniors who are smartphone users (46.7% women). Participants were randomly assigned to either a preannouncement or website news condition. Participants assigned to the preannouncement condition were presented with a message that said, 'This message is from the homepage of M company: Today M Company is happy to announce that we will launch our new smartphone in the near future. The preannouncement picture of the new smartphone is shown below.' At the bottom of the message, participants saw a picture of a smartphone from the Internet. Participants assigned to the online news condition saw a message that said, 'This message is from Sina - one of the most famous new websites in China: An employee from M company says that M company probably will launch a smartphone in the near future. A leaked picture of the new smartphone is shown below.' At the bottom of the message, participants saw the same picture used in Scenario 1 (see Table 1). After participants read the

Table 1. Descriptive scenario designs for pilot study.

a1 Official preannouncement	a2 Announcement from News Website
This message is from the homepage of M company: Today M Company is happy to officially announce that we will launch our new smartphone in the near future. We are providing an official preannouncement picture of the new smartphone below.	This message is from Sina.com: An employee from M company says that M company will probably launch a smartphone in the near future. Below is a leaked picture of the new smartphone.

messages, they completed a questionnaire including basic information about the different scenario messages and pictures, questions about their eWOM, and their demographics.

Engagement in eWOM was operationalized with three behaviours: opinion seeking, opinion giving, and opinion passing (Flynn, Goldsmith, and Eastman 1996; Sun et al. 2006) and measured with three items adopted from Chu and Kim (2011): 'I like to get opinions through the Internet about new products before I buy them,' 'I like to share my opinions about new products on the Internet,' 'I like to pass along interesting information about new products on SNS.' For each item, participants reported their eWOM on a five-point scale (1 = *not very likely*, 5 = *very likely*). We used the same items to measure eWOM for Studies 2 and 3. To check whether NPP message manipulation was successful, we ran ANOVA on opinions about whether the message was official. The analysis revealed a significant difference: participants in the NPP message condition perceived the message to be more official ($M = 3.81$) than did those in the online news condition ($M = 2.20$; $F(1, 58) = 20.52$; $p < .001$), suggesting that the NPP message manipulation was successful.

Next, we tested whether the conditions were significantly different. The results showed significant eWOM differences between the two scenarios ($F(1, 58) = 9.88$, $p < 0.01$). The mean score of eWOM on the NPP condition was higher than the score on the website news condition ($M_{NPP} = 4.33 > M_{News} = 3.58$). The study supports the prediction that NPP can generate more eWOM. Official announcements about launching a new product can attract more attention than other types of information. Observers are more likely to share opinions and messages about NPP information. Based on this evidence, we tested the hypotheses in Studies 1, 2, and 3.

Study 1: The effects of NPP message clarity on eWOM

Method

Sample. Participants were 60 Chinese business school juniors who are smartphone users (53.3% women). China, which has a population of 600 million, is home to the world's largest Internet-using population. Most Internet users are 20 to 29-year-old (30.3%) (Statista

2016). Most social network users are 19 to 30-years-old (49%) (Websitehub 2015). Although we did not actually measure SNS usage rates, we felt that the age range made it appropriate to sample university students. Participants reported spending an average of US$225 when they purchased their smartphones. The most popular smartphone brands were Samsung, Apple, Xiaomi, and Huawei; 63.3% were planning to change their smartphones within one year; 78.9% planned to spend an average of US$277 on their next purchase.

Procedure. Study 1 included two scenarios testing the effects of clarity in NPP messages. Participants were randomly assigned to the high clarity or low clarity message condition. Both conditions were introduced with: 'This message is from the homepage of M company: Today M Company is happy to officially announce that we will launch our new smartphone in the near future. The official preannouncement picture is shown below.' Below the message, we provided a clear illustration and message or an unclear illustration and message. The high clarity message provided an image of the smartphone clearly showing shape, size, and color, along with details regarding CPU, ram, screen, and camera pixels. The low-clarity less-detailed message condition showed the F6 logo against a sea of stars, without concretely picturing the product or providing product information. After participants viewed the messages, they completed a questionnaire including demographics, basic information about the scenario messages and pictures, and questions about their eWOM (see Table 2).

Dependent measures. Engagement in eWOM was operationalized according to three behaviors: opinion seeking, opinion giving, and opinion passing (Flynn, Goldsmith, and Eastman 1996; Sun et al. 2006) and measured with three items adopted from Chu and Kim (2011): 'I like to get opinions through the Internet about new products before I buy them,' 'I like to share my opinions about new products on the Internet,' 'I like to pass along interesting information about new products on SNS.' For each item, participants reported their eWOM on a five-point scale (1 = *not very likely*, 5 = *very likely*). We used the same items to measure eWOM for the rest of the studies.

Table 2. Descriptive scenario designs for Study 1.

a1 High clarity NPP	a2 Low clarity NPP
This message is from the homepage of M company: Today M Company is happy to officially announce that we will launch our new smartphone in the near future. The official preannouncement picture of the new smartphone is shown below.	This message is from the homepage of M company: Today M Company is happy to officially announce that we will launch our new smartphone in the near future. The official preannouncement picture of the new smartphone is shown below.

Table 3. Test of NPP message clarity.

	Scenario	Mean	S.D.	F Value	p Value
eWOM	High clarity	4.10	.84	14.33***	.000
	Low clarity	3.17	.88		

***$p < .001$.

Manipulation check. To validate the manipulation of the clarity of NPP message, we conducted a separate pretest in which 60 business school graduate students received the NPP manipulation described above and reported the extent to which they thought that the two NPP messages were 1 = *very unclear,* 5 = *very clear.* The analysis revealed a significant difference between the two messages: Participants in the high clarity condition reported significantly higher message clarity (M = 3.05) than did those in the low clarity condition (M = 2.15; $F(1, 58) = 13.70$; $p < .001$), which confirmed successful manipulation of the message.

Results and discussion

ANOVA results showed significant differences in eWOM between the two NPP messages ($F(1, 58) = 14.33$, $p < 0.001$). The mean score of eWOM on the high clarity NPP condition was higher than the score for the low clarity NPP condition ($M_{high\ clarity} = 4.10 > M_{low\ clarity} = 3.17$) (see Table 3).

The results support the prediction that the clarity of NPP messages can significantly increase eWOM. More specific information about new products can reduce ambiguity and uncertainty, which leads to higher eWOM. Thus, Hypothesis 1 is supported. In Studies 2 and 3, we examined whether brand characteristics will moderate the effects of NPP message clarity.

Study 2: Moderating effects of brand familiarity

Method

Sample. Participating in Study 2 were 120 business school junior undergraduates (54.2% women) who were smartphone users. On average, they reported spending US$214 when they purchased their smartphones. The most popular smartphone brands were Samsung, Apple, Xiaomi, and Huawei; 56.7% were planning to buy another smartphone within one year; 70.6% planned to spend an average of US$287 on their next purchase.

Procedure. Study 2 included four scenarios using a 2 (NPP message clarity: high clarity vs. low clarity) × 2 (Brand familiarity: high vs. low) between-subjects design. We randomly assigned participants to each treatment condition. *Huawei* was selected as the familiar brand; and *Bihee* as the unfamiliar brand, following Sundaram and Webster (1999). Participants were asked to indicate their level of familiarity with the brand name to which they were exposed,

Table 4. The effects of NPP message clarity and brand familiarity on eWOM.

Dependent variable	Factors	Mean square	df	F	P
eWOM	NPP message	46.875	1	97.04***	0.000
	Brand familiarity	16.875	1	34.94***	0.000
	Clarity*brand familiarity	8.01	1	16.58***	0.000

***$p < .001$.

using a 5-point scale: 1 = *not familiar* and 5 = *very familiar*. The analysis indicated significant differences ($F(1, 118) = 196.54, p < .001$) between familiar ($M = 3.77$) and unfamiliar ($M = 1.73$) brands, which confirmed that brand familiarity was adequately manipulated.

As in Study 1, the high clarity NPP message condition showed a more concrete picture displaying the smartphone's shape, size, and colour, and included more specific information such as CPU, ram, screen, and camera pixels. In the low clarity NPP message condition, the phone was not pictured and specific product details were not provided. The manipulation check revealed a significant difference in NPP message clarity: the high clarity NPP condition showed significantly higher clarity ($M = 3.71$) than the low clarity condition ($M = 1.75$; $F(1, 118) = 54.98$; $p < .001$), indicating that the NPP message clarity was adequately manipulated. After participants viewed the messages, they completed a questionnaire regarding their eWOM behaviour and demographics.

Results and discussion

To test the effects of message clarity and brand familiarity as a moderator, we used a 2 (high vs. low clarity NPP message) × 2 (familiar vs. unfamiliar brand) ANOVA on eWOM. NPP message ($F(1,118) = 97.04, p < .001$) and brand familiarity ($F(1,118) = 34.94, p < .001$) had main effects on eWOM. NPP message and brand familiarity showed significant interaction ($F(1,116) = 16.58, p < .001$), as Table 4 shows. For the familiar brand, high clarity NPP message generated significantly higher eWOM than the low clarity NPP ($M_{high\ clarity\_fam} = 4.20$ vs. $M_{low\ clarity\_fam}$ 3.46). For the unfamiliar brand condition, the high clarity NPP message also generated significantly higher eWOM than the low clarity NPP ($M_{high\ clarity\_unfam} = 3.96$ vs. $M_{low\ clarity\_unfam} = 2.20$). However, the unfamiliar brand showed a much steeper increase than the familiar brand between the message conditions, as Figure 2 shows.

The Study 2 results showed that NPP message clarity more strongly influenced eWOM for the unfamiliar brand than it did for the familiar brand. Participants in the unfamiliar brand

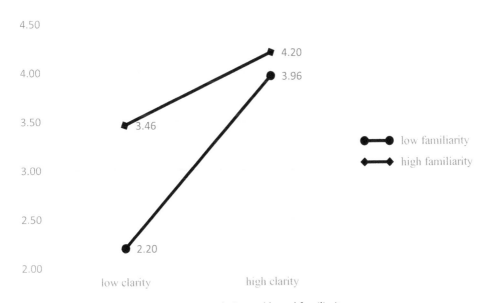

Figure 2. Interaction between NPP message clarity and brand familiarity.

with low message clarity condition indicated having the lowest eWOM intentions. However, eWOM intentions for unfamiliar brands were strongly boosted when the NPP message had high clarity. Although the familiar brand showed a similar pattern in the effects of message clarity, eWOM was less likely for the unfamiliar brand. Thus, Hypothesis 2 is supported.

Study 3: Moderating effects of brand preference

Method

Sample. Participating in Study 3 were 120 business school undergraduate seniors (53.3% women) who were smartphone users. On average, they reported spending US$244 when they bought their smartphones. The most popular smartphone brands were Apple, Samsung, Xiaomi, and Huawei; 44.2% were planning to buy new smartphones within one year; 90.6% planned to spend an average of US$304 on their next purchase.

Procedure. Study 3 applied four scenarios using a 2 (high vs. low NPP message clarity) × 2 (high vs. low brand preference) between-subjects design. Following Liu and Smeesters's (2010) procedure, participants were randomly assigned to each treatment condition. Participants assigned to the preferred brand condition were asked to write down their preferred brand name and describe their reasons for the preference. Participants in the non-preferred brand condition were asked to write down a brand name they did not prefer and describe their reasons. As in Studies 1 and 2, the details of NPP messages were manipulated in pictorial and textual product information.

After participants were exposed to the NPP messages, they completed a questionnaire about their demographics, their preference for the brand, and their eWOM intentions.

To manipulate the brand preference in the NPP message, participants were asked to respond to two items: 'I prefer this smartphone brand over others,' and 'It is likely that I will buy this brand when I make my next purchase,' using a 5-point scale (1 = *strongly disagree* and 5 = *strongly agree*). The questions were adopted from Sääksjärvi and Samiee (2011) (alpha = .91). The analysis indicated significant difference ($F(1, 118) = 108.14, p < .001$) between the highly preferred brand ($M = 3.76$) and the less-preferred brand ($M = 1.40$), suggesting that the level of brand preference was adequately manipulated. As in prior experiments, we confirmed a significant difference ($F(1, 118) = 13.13, p < .01$) in the NPP messages between high clarity ($M = 4.29$) versus low clarity ($M = 2.70$) conditions.

Results

To test the moderation effects of brand preference, we used a 2 (high vs. low message clarity) × 2 (high vs. low brand preference) ANOVA on eWOM. Results showed that message clarity ($F(1, 118) = 31.57, p < .001$) and brand preference ($F(1, 118) = 43.50, p < .001$) had main effects on eWOM. Message clarity also interacted significantly with brand

Table 5. The effects of NPP message clarity and brand preference on eWOM.

Dependent variable	Factors	Mean square	df	F	P
eWOM	NPP message	17.63	1	31.567***	.000
	Brand preference	24.30	1	43.500***	.000
	NPP message*brand preference	5.63	1	10.084***	.002

***$p < .001$

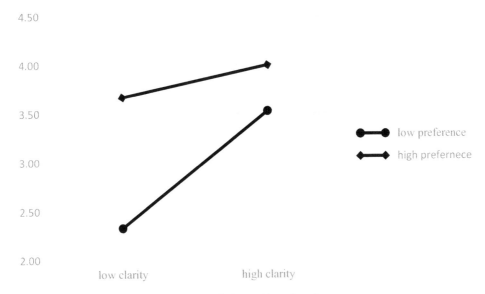

Figure 3. Interaction between NPP message clarity and brand preference.

preference ($F(1, 116) = 10.08$, $p < .01$), as Table 5 shows. For the highly preferred brand, participants indicated significantly higher eWOM intentions for high rather than low message clarity ($M_{\text{high clarity\_pref}} = 4.00$ vs. $M_{\text{low clarity\_pref}} = 3.67$). For the less-preferred brand, high clarity NPP showed much higher eWOM than the low clarity NPP ($M_{\text{high clarity\_n pref}} = 3.53$ vs. $M_{\text{low clarity\_n pref}} = 2.33$). However, the difference between the message clarity conditions was much higher when participants had low rather than high brand preference, as shown in Figure 3.

Message clarity had a stronger influence on eWOM for the less-preferred brand than for the preferred brand. When participants had low brand preference and the NPP message had low clarity, eWOM intentions were the lowest. Comparatively, in the low brand preference condition, when participants viewed NPP messages of high clarity, their eWOM intentions rose strongly, higher than they did in the high brand preference condition. Thus, Hypothesis 3 is supported.

Discussion

NPPs are one of the most widely used signalling tools in which companies diffuse product information to target groups before launching new products. In this research, we study the effects of NPP message clarity on eWOM interacting with brand characteristics.

In a pilot test, we directly compare NPP with online news and find that formal NPP messages do, in fact, more strongly motivate consumers to spread eWOM. In Study 1, we find that consumers are more likely to disseminate eWOM when NPP messages are highly clear and provide more product details. In Studies 2 and 3, we show that brand characteristics interact with NPP message clarity to affect eWOM intentions.

Our results make several contributions to the extant literature on NPPs and eWOM. First, we provide a potential benchmark empirical study examining how NPP message features affect eWOM. Signalling theory has viewed NPP as an important strategy for alleviating uncertainty surrounding products, firms, or markets (Gatignon and Robertson 1991; Schatzel and Calantone 2006), but the influences on eWOM are less understood. We address this issue by developing new insights into the role of NPPs in determining whether consumers will share promotional messages. We find that exposure to NPP messages indeed influences eWOM behaviour, but the relative effects of message clarity depend on brand characteristics.

Second, we extend research on the effects of NPP message clarity on e-WOM behaviour. Prior signalling literature has suggested that message clarity improves shareholder value and brand equity by making brand information more credible (e.g. Erdem and Swait, 1998; Sorescu, Shankar, and Kushwaha 2007). We contribute to signalling theory by showing that consumers will perceive that NPP messages providing more specific product details indicating high quality are then encouraged to spread positive eWOM.

Third, we show that brand familiarity and brand preference have moderation effects. That is, underdog brands stand to gain more benefits from highly clear NPP messages. Although our study shows that brand strength favourably affects e-WOM, we also find that NPP message clarity has stronger influence for less-familiar or less-preferred brands, probably by lowering uncertainty or perceived risk. The results can be interpreted also from the perspective of the theory of preference-decision consistency. Consumers will readily share NPP messages about well-known or preferred brands, but for unknown or less-preferred brands, highly clear NPP messages may cause them to feel that external brand information being signalled is inconsistent with their brand knowledge. They will have greater intentions to create eWOM when their experiences are inconsistent with others' postings (Shin et al. 2014), which may indicate why brand characteristics modify the effects of NPPs.

Our studies provide several managerial insights for preannouncement strategies. First, new product development strategy includes decisions about which opportunities to pursue, how much to invest in R&D, and what messages to communicate (Ofek and Turut 2013). Our findings indicate that firms should include NPPs as a strategy in their new product launching packages as the first step to alert customers to upcoming events, persuade them to postpone making purchases until the new product comes into the market, and motivate them to express positive sentiments when they share brand information. Second, designers of NPP messages must consider the level of message clarity and brand characteristics. Though clear messages with specific product details are beneficial for both famous and unknown brands, the signalling effects of NPPs for product quality appear to be much stronger for unknown or less-preferred brands.

The studies provide strategic approaches for NPP communication, but a few limitations should be noted. First, we manipulated the text information in the pilot test only, but we might have established a clearer distinction if we had included brand components. For example, an official announcement can be designed with a specific brand mark, whereas unofficial news can include the logo of the news media. Second, we focused on message clarity and brand characteristics as they affect NPP. However, future research can consider a broader range of additional message factors such as message consistency, media channels to deliver NPP signals, message format (e.g. visual vs. verbal message; emotional vs. rational appeals), and product types (e.g. high vs. low product involvement). For example, senders' construal level and temporal distance are known to cause the visual elements of an advertising message to have different effects on eWOM (Choi, Seo, and Yoon 2017; Trope and Liberman 2010). Furthermore, product types such as high versus low involvement or search versus experiential goods can moderate NPP messages. Therefore, the study should be replicated across various message formats (e.g. pictorial vs. text-oriented messages) and product categories (e.g. films, software, automobiles, etc.). Third, we used real brands to examine the effects of brand characteristics. However, possible confounds could influence WOM and might be considered in the design of stimuli (e.g. prior knowledge of the brand and the brand's country of origin). We manipulated message clarity according to picture concreteness and the level of product details in the text message. Because verbal and visual information can have different effects on the clarity of NPPs, future research should test how these components interact to influence eWOM. More important, when consumers consider non-preferred brands, they may tilt toward negative eWOM (e.g. Park and Lee 2009). Thus, future research should consider positive and negative eWOM sentiments. Finally, though college students are the major Internet and SNS users, different demographic segments may show different eWOM behavior. Therefore, future research should include more diverse demographic groups.

Acknowledgments

The authors thank the editor, guest editors, and anonymous reviewers for their helpful comments during the review process.

Disclosure statement

No potential conflict of interest was reported by the author(s).

Funding

National Natural Science Foundation of China [grant number 71672027].

References

Aaker, D.A. 2002. Building strong brands. *Brandweek* 24, no. 2: 115–8.
Alba, J.W., and J.W. Hutchinson. 1987. Dimensions of consumer expertise. *Journal of Consumer Research* 13, no. 4: 411–54.
Bayus, B.L., S. Jain, and A.G. Rao. 2001. Truth or consequences: An analysis of vaporware and new product announcements. *Journal of Marketing Research* 38, no. 1: 3–13.
Biswas, A. 1992. The moderating role of brand familiarity in reference price perceptions. *Journal of Business Research* 25, no. 3: 251–62.
Campbell, M.C., and K.L. Keller. 2003. Brand familiarity and advertising repetition effects. *Journal of Consumer Research* 30, no. 2: 292–304.

Chen, C., and Y. Chang. 2008. Airline brand equity, brand preference, and purchase intentions–The moderating effects of switching costs. *Journal of Air Transport Management* 14, no. 1: 40–2.

Chen, C., and V. Wong. 2012. Design and delivery of new product preannouncement messages. *Journal of Marketing Theory & Practice* 20, no. 2: 203–22.

Choe, P., and Y. Zhao. 2013. The influence of airline brand on purchase intention of air tickets in China. *Industrial Engineering and Management Systems* 12, no. 2: 34–46.

Choi, Y., Y. Seo, and S. Yoon. 2017. E-WOM messaging on social media: Social ties, temporal distance, and message concreteness, *Internet Research* 27, no. 3: 495–505.

Chu, S.C., and Y. Kim. 2011. Determinants of consumer engagement in electronic word-of-mouth (eWOM) in social networking sites. *International Journal of Advertising* 30, no. 1: 47–75.

Daft, R.L., R.H. Lengel, and L.K. Trevino. 1987. Message equivocality, media selection, and manager performance: Implication for information systems. *MIS Quarterly* 11, no. 3: 355–66.

De Angelis, M., A. Nonezzi, A.M. Peluso, D.D. Rucker, and M. Costabile. 2012. On braggarts and gossips: A self-enhancement account of word-of-mouth generation and transmission. *Journal of Marketing Research* 54: 551–63.

Donath, J. 2007. Signals in social supernets. *Journal of Computer-Mediated Communication* 13, no. 1: 231–51.

Duhan, D.F., S.D. Johnson, J.B. Wilcox, and G.D. Harrell. 1997. Influence on consumer use of word-of-mouth recommendation sources. 25, no. 4: 283–95.

Eliashberg, J. and T.S. Robertson. 1988. New production preannouncing behavior: A market signaling study. *Journal of Marketing Research* 25, no. 8: 282–92.

Erdem, T., and J. Swait. 1998. Brand equity as a signaling phenomenon. *Journal of Consumer Psychology* 7, no. 2: 131–57.

Fill, C. 1995. *Marketing communications: Frameworks, theories, and applications.* Hemel Hempstead: Prentice Hall International.

Flynn, L.R., R.E. Goldsmith, and J.K. Eastman. 1996. Opinion leaders and opinion seekers: two new measurements scales. *Journal of the Academy of Marketing Science* 24, no. 2: 137–47.

Garcia-Granero, A., O. Llopis, A. Fernandez-Mesa, and J. Alegre. 2015. Unraveling the link between managerial risk-taking and innovation: The mediating role of a risk-taking climate. *Journal of Business Research* 68, no. 5: 1094–104.

Gatignon, H., and T.S. Robertson. 1991. An exchange theory model of interpersonal communication. *Advances in Consumer Research* 13, no. 1: 534–8.

Gilly, M.C., J.L. Graham, M.F. Wolfinbarger, and L.J. Yale. 1998. A dyadic study of interpersonal information search. *Journal of the Academy of Marketing Science* 26, no. 2: 83–100.

Goldsmith, R.E., L.R. Flynn, and E.B. Goldsmith. 2003. Innovative consumers and market mavens. *Journal of Marketing Theory and Practice* 11, no. 4: 54–64.

Goldsmith, R.E., and D. Horowitz. 2006. Measuring motivations for online opinion seeking. *Journal of Interactive Advertising* 6, no. 2: 1–16.

Grewal, D., G. Iyer, and M. Levy. 2004. Internet retailing: Enablers, limiters, and market consequences. *Journal of Business Research* 57, no. 7: 703–13.

Haenlein, M., and B. Libai. 2013. Targeting revenue leaders for a new product. *Journal of Marketing* 77, no. 3: 65–80.

Heil, O.P. and T. Roberson 1991. Toward a theory of competitive market signaling: A research agenda. *Strategic Management Journal* 12, no. 6: 403–18.

Heil, O.P. and R.G. Walters 1993. Explaining competitive reactions to new products: An empirical signaling study. *Product Innovation Management* 10, no. 1: 53–65.

Hennig-Thurau, T., K.P. Gwinner, G. Walsh, and D.D. Gremler 2004. Electronic word-of-mouth via consumer-opinion platforms: What motivates consumers to articulate themselves on the internet? *Journal of Interactive Marketing* 18, no. 1: 38–52.

Hoyer, W.D., and S.P. Brown. 1990. Effects of brand awareness on choice for a common, repeat-purchase product. *Journal of Consumer Research* 17, no. 2: 141–8.

Jamal, A., and M. Al-Marri. 2007. Exploring the effects of self-image congruence and brand preference on satisfaction: The role of expertise. *Journal of Marketing Management* 23, no. 7–8: 613–29.

Jamal, A., and M.M.H. Goode. 2001. Consumers and brands: A study of the impact of self-image congruence on brand preference and satisfaction. *Marketing Intelligence & Planning* 19, no. 7: 482–92.

Khasawneh, M.A., and A. Shuhaiber. 2013. A comprehensive model of factors influencing consumer attitude towards and acceptance of SMS advertising: an empirical investigation in Jordan. *International Journal of Sales & Marketing Management Research and Development* 3, no. 2: 1–22.

Kim, B.H., S. Han, and S. Yoon. 2010. Advertising creativity in Korea. *Journal of Advertising* 39, no. 2: 93–108.

King, R.A., P. Racherla, and V.D. Bush. 2014. What we know and don't know about online word-of-mouth: A review and synthesis of the literature. *Journal of Interactive Marketing* 28, no. 3: 167–83.

Kirmani, A., and A. Rao. 2000. No pain, no gain: A critical review of the literature on signaling unobservable product quality, *Journal of Marketing* 64, no. 2: 66–79.

Kohli, C. 1999. Signaling new product introductions: A framework explaining the timing of preannouncements. *Journal of Business Research* 46, no. 1: 45–56.

Koku, P.S. 1997. Corporate name change signaling in the services industry. *Journal of Services Marketing* 11, no. 6: 392–408.

Lau, G.T., and S. Ng. 2001. Individual and situational factors influencing negative word-of-mouth behaviour. *Canadian Journal of Administrative Science* 18, no. 3: 163–78.

Lee, M., and S. Youn. 2009. Electronic word of mouth (eWOM) – how eWOM platforms influence consumer product judgment. *International Journal of Advertising* 28, no. 3: 473–99.

Lilly, B., and H.S. Krishnan. 1996. Consumer responses to new product announcements: A conceptual framework. In *1996 winter educators' conference, marketing theory and applications*, ed. E.A. Blair and W. A Kamakura, 56–62. Chicago: American Marketing Association.

Lilly, B., and R. Walters. 1997. Toward a model of new product preannouncement timing. *Journal of Product Innovation Management* 14, no. 1: 4–20.

Lilly, B., and R. Walters. 2000. An exploratory examination of retaliatory preannouncing. *Journal of Marketing Theory & Practice* 8, no. 4: 1–9.

Liu, J., and D. Smeesters. 2010. Have you seen the news today? The effect of death-related media contexts on brand preference. *Journal of Marketing Research* 47, no. 2: 251–62.

Mau, G., G. Silberer, and C. Constien. 2008. Communicating brands playfully: Effects of in-game advertising for familiar and unfamiliar brands. *International Journal of Advertising* 27, no. 5: 827–51.

Mavlanova, T., R. Benhunan-Fich, and M. Koufaris. 2012. Signaling theory and information asymmetry in online commerce. *Information & Management* 49, no. 5: 240–7.

Mishra D.P., and H.S. Bhabra. 2001. Assessing the economic worth of new product pre- announcement signals: Theory and empirical evidence. *Journal of Product & Brand Management* 10, no. 2: 75–93(19).

Ofek, E., and O. Turut. 2013. Vaporware, suddenware, and trueware: New product preaanoucement under market uncertainty. *Marketing Science* 32, no. 2: 342–55.

Park, C., and T.M. Lee. 2009. Information direction, website reputation and eWOM effect: A moderating role of product type. *Journal of Business Research* 62, no. 1: 61–7.

Rabino, S., and T.E. Moore. 1989. Managing new product announcements in the computer industry. *Industrial Marketing Management* 18, no. 1: 35–43.

Ramsoy, T.Z. and M. Skov. 2014. Brand preference affects the threshold for perceptual awareness. *Journal of Consumer Behavior* 1, no. 1: 1–8.

Robertson, T.S., and T. Rymon. 1995. New product announcement signals and incumbent reactions. *Journal of Marketing* 59, no. 3: 1–15.

Sääksjärvi, M., and S. Samiee. 2011. Relationships among brand identity, band image and brand preference: Differences between cyber and extension retail brands over time. *Journal of Interactive Marketing* 25, no. 3: 167–77.

Schatzel, K., and R. Calantone. 2006. Creating market anticipation: An exploratory examination of the effect of preannouncement behavior on a new product's launch. *Journal of the Academy of Marketing Science* 34, no. 3: 357–66.

Sen, S., and D. Lerman. 2007. Why are you telling me this? An examination into negative consumer reviews on the web. *Journal of Interactive Marketing* 21, no. 4: 76–94.

Shin, D., J.H. Song, and A. Biswas. 2014. Electronic word-of-mouth (eWOM) generation in new media platform: The role of regulatory focus and collective dissonance. *Marketing Letters* 25, no. 2: 153–65.

Sorescu, A., V. Shankar, and T. Kushwaha. 2007. New product preannouncements and shareholder value: Don't make promises you can't keep. *Journal of Marketing Research* 44, no. 3: 468–89.

Statista. 2016. Breakdown of Internet users in China from December 2014. December 2016, by age, [https://www.statista.com/statistics/265150/internet-users-in-china-by-age/].

Su, M., and V.R. Rao. 2010. New product preannouncement as a signaling strategy: An audience-specific review and analysis. *Journal of Product Innovation Management* 27, no. 5: 658–72.

Su, M., and V.R. Rao. 2011. Timing decisions of new product preannouncement and launch with competition. *International Journal of Production Economics* 129, no. 1: 51–64.

Sun, T., S. Youn, G. Wu, and M. Kuntaraporn. 2006. Online word-of-mouth (or mouse): An exploration of its antecedents and consequences. *Journal of Computer-Mediated Communication* 11, no. 4: 1104–27.

Sundaram, D.S., and C. Webster. 1999. The role of brand familiarity on the impact of word-of-mouth communication on brand evaluations. *Advances in Consumer Research* 26, no. 1: 664.

Taylor, C.R., G.R. Franke, and H. Bang. 2006. Use and effectiveness of billboards perspectives from selective-perception theory and retail-gravity models. *Journal of Advertising* 35, no. 4: 21–34.

Trope, Y., and N. Liberman. 2010. Construal-level theory of psychological distance. *Psychological Review* 117, no. 2: 440–63.

Websitehub. 2015. 2015 Chinese social media statistics and trends inforgraphic. https://makeawebsitehub.com/chinese-social-media-statistics/

Westbrook, R.A. 1987. Product/consumption-based affective responses and post purchase processes. *Journal of Marketing Research* 24, no. 3: 258–70.

Wind, J., and V. Mahajan. 1987. Marketing hype: a new perspective for new product research and introduction. *Journal of Product Innovation Management* 4, no. 1: 43–9.

Yap, K.B., B. Soetarto, and J.C. Sweeney. 2013. The relationship between electronic word-of-mouth motivations and message characteristics: The senders' perspective. *Australasian Marketing Journal* 21, no. 1: 66–74.

The power of e-WOM using the hashtag: focusing on SNS advertising of SPA brands

Jiye Shin, Heeju Chae and Eunju Ko

ABSTRACT
The dissemination and utilization of personal mobile social network services (SNS) have created new trends. The effects of the increase in the use of the hashtag are especially significant for brands that seek to use electronic word-of-mouth (e-WOM) in the rapidly changing environment of SNS marketing strategies, as it supplements their quick merchandise turnover rates. This study aims to examine the advertising effects of marketing activities (product, place, promotion, and price) and the hashtag used in the SNS advertising of SPA brands. The research collects words related to SPA brands through the analysis of the data of SNS utilized over six months. To test the hypothesis, an experimental design of 4 (product/place/promotion/price) × 2 (hashtag/no hashtag) was carried out. The results showed that the differences for various types of SNS fashion marketing, broadening the scope of existing research studies that focus on the impact of SNS in the marketing environment.

1. Introduction

The recent dissemination of smartphones and the introduction of Web 3.0 have enabled users to easily create and share content through social network services (SNS), one of the most effective communication channels. In particular, the ubiquitous use of smartphones have given users frequent and rapid access to SNS, so that usage rate increased from 8.6% in 2013 to 39.9% in 2014 (Kim 2015).

Younger consumers often require immediately available information, seek products and services that satisfy their personal desires, and prefer business models that gratify their inner needs (Nam, Son, and Lee 2015). SNS has become popular because it is a way to communicate conveniently, free from time and spatial limitations. Consequently, the hashtag # symbol has emerged as a marketing tool allowing consumers to easily express their values (Oh 2015), most frequently through SNS platforms. Hashtag symbols precede keywords, which then can be searched for a collective display of related information (Chae, Ko, and Han 2015). Especially in SPA fast-fashion brands, where various promotions and special offers accompany rapid merchandise turnover rates, the hashtag is used as an

SNS marketing strategy to maintain close contact with consumers (Chae, Ko, and Han 2015). In SNS environments, SPA brands may expect to promote effective e-WOM by sparking consumer interest and participation using the hashtag, thus creating a low-cost, high-efficiency advertising effect. This outlook calls for research to objectively verify the businesses effectiveness of the hashtag, to see what positive effects accrue from advertisements using hashtags, and whether advertising effects are substantially influenced. Nonetheless, little research has been conducted to identify factors of SPA brand marketing activities regarding SNS and hashtag usage. Therefore, in this research, we examine the advertising effects of the hashtag, based on Ducoffe's (1996) web advertising model and preceding research on e-WOM (Chatterjee 2011; Chu and Kim 2011).

We conducted this study to analyze SNS user-created big data related to SPA brands. We conducted in-depth interviews to derive keywords surrounding SPA brands and further categorized the keywords into marketing activities (product/brand/promotion/price). Second, we examined differences among informativeness, enjoyment, interactivity, attitudes toward advertising, and e-WOM based on the hashtag (hashtag/no hashtag), and other SPA brand marketing activities in SNS advertisements. Third, we investigated the influence that informativeness, enjoyment, and interactivity have on attitudes toward advertising and e-WOM within hashtag advertisements for SPA brands. Moreover, by analyzing the substantive advertising effects of keywords under different marketing activities, derived from the analysis of consumer-created big data, we provide data to rapidly growing SPA brands needing effective SNS advertising strategies.

2. Conceptual framework and hypotheses

2.1. Hashtag marketing; new paradigm of e-WOM

Potential customers directly and indirectly express extensive vivid descriptions, opinions, comments, and emotions about products through SNS (Ko et al. 2015). Excessive indiscrete data causes fatigue, avoidance, and other negative effects, so consumers prefer information that is specifically related to their situations. In particular, the hashtag is considered a true source for ascertaining the 'community of taste' (Shin 2016). The hashtag is a Twitter function that uses '#' in front of a particular word representing the topic. Other services including Facebook, Instagram, Google Plus, Tumblr, Flickr, and YouTube support hashtags. Recently, Korea's leading IT corporations such as Kakao and Naver have introduced the hashtag function (Oh 2015) to describe products, brands, and fashions, users' locations, or buzzwords that express consumer feelings. These brands mainly target younger consumers in their 20s and 30s according to their current trends, desires, and interests (Oh 2015). Fashion brands are now using hashtags for their simple and swift characteristics: they are so easy and useful for searching and collecting information that draw active consumer participation and attention, which in turn produces e-WOM and popularity. For example, DKNY and Marc by Marc Jacobs used hashtags to draw attention to their promotional model, while Calvin Klein, Michael Kors, Adidas, Topshop, and various other fashion brands use hashtag to connect with consumers (Jones 2014). Hashtags allow consumers to become more aware of and interested in brands. Positive reactions represented by 'likes' or 're-hashtags' can grab attention and stimulate curiosity about forthcoming products. Aligned with this trend, Instagram created a shopping functionality within the platform allowing users to reach the retailer and buy the product simply by tapping on a hashtag picture, with one click (Chaykowski 2016).

Thus, hashtags are a form of e-WOM. In terms of SPA brands, with their rapid merchandise-turnover rates, promotions, and special offers, hashtags should allow more effective real-time communication with consumers at relatively low prices. Indeed, hashtags have been shown to significantly affect e-WOM and the number of Twitter retweets (Lee and Jang 2013). A case analysis and typology combining social media and gamification – with a focus on Facebook's fan page and Instagram's hashtag marketing (Chae, Ko, and Han 2015, Moon, Lee, and Park 2015) – showed motivations behind using hashtags in the growing image-based SNS. Interest sharing, social interaction, ease of use, and enjoyment were shown to affect social participation and overall brand assets.

Fashion items are especially likely to draw consumer attention to real-time information updates or trends (Nam, Son, and Lee 2015). Luxury brands such as Burberry and Louis Vuitton and global SPA brands including UNIQLO, H&M, and ZARA are examples of brands using SNS channels such as Facebook, Twitter, Instagram, Pinterest, and You-Tube for brand promotion and communication (Chae, Shin, and Ko 2015). Online shopping platforms generate actual sales (Ko et al. 2014). Thus, by using rapidly spreading SNS, brands can increase e-WOM. Additionally, they can utilize functions such as *retweet*, *like*, *share*, and *hashtags* to maximize emotional interaction with consumers (Park and Cho 2014).

Therefore, in this research, we consider the dimensions of hashtag marketing activities of SPA brands, including product, place, promotion, and price, and then examine the differences in their advertising effects.

2.2. Attitude toward SNS advertising

Consumers' subjective valuation of the relative value or effectiveness of an advertisement determines *advertising value*, a potential measurement of advertising market orientation (Ducoffe 1996). Ducoffe's (1996) web advertising effectiveness model includes entertainment, informativeness, and irritation antecedents that influence attitudes toward advertising. Among the antecedents, *enjoyment* has the most direct effect.

'For advertisers, this implies that by selecting media carefully that fit the communication task at hand and media vehicles that accurately target the most interested potential customers; they can enhance the advertising value' (Ducoffe 1995). Therefore, SNS advertising of SPA brands through hashtags potentially adds advertising value. Consequently, our research focuses on the antecedents of advertising value of SNS advertising. Taylor, Lewin, and Strutton (2011) examined SNS advertising value and attitudes toward advertising and suggested a model of consumer response to SNS advertisements. That is, entertainment and informativeness factors significantly affect attitudes toward SNS advertisements, with an emphasis on the enjoyment factor.

Online advertisements are the typical channels through which consumers may encounter either push advertising in which receivers are spammed and irritated, or bilateral pull consumer-created and shared SNS advertisements (Sung and Cho 2011). In this research, we examine only pull-type SNS advertising through KakaoTalk, Instagram, and Facebook, where users may subscribe to their brands of interest. We also focus on the enjoyment and informativeness factors.

Quantitative research has examined how the orientation and type of Twitter messages affect e-WOM, and has proven that hashtags significantly affect the number and speed of retweets (Lee and Jang 2013). Thus, users can use hashtags as an archive allowing them to find desired information through a keyword search (Moon, Lee, and Park 2015). Thus, we hypothesize that hashtags will affect advertising value.

H1: Perceived informativeness will differ by use of hashtag according to marketing activities.

H2: Perceived enjoyment will differ by use of hashtag according to marketing activities.

2.3. Interactivity

The development of interactive media, such as the Internet and mobile networks, now gives informational control to consumers, not exclusively to producers (Bezjian-Avery, Calder, and Iacobucci 1998). This signifies that interactivity could become an important variable. Research by Lee and Cho (2011) confirms that higher interactivity leads to a better attitude towards advertising, participation intention, and e-WOM intention regarding interactive advertisements. Notably within the SNS environment, the dissemination of the hashtag is a means to speed up interactivity; in the marketing field, maintaining a sound relationship with the customers through interactivity is one of the most important assets of any business (Duncan and Moriarty 1998; Kennedy, Ferrell, and LeClair 2001). In the research by Chae, Shin, and Ko (2015), the SPA brands' interactivity in SNS is an important determinant of communication and relationship maintenance with customers. Due to SNS's characteristics as a one-to-many medium, continuous research on the interactivity factors is being conducted. For shopping situations, such characteristics retain an even greater significance (Tang and Solomon 1998). Based on this research, the following hypothesis has been set to examine the effect that the type of advertising methods derived from SNS advertising.

H3: There will be a difference in the interactivity in SNS advertising of SPA brands, based on the use of the hashtag and marketing activities.

2.4. Advertising effects

2.4.1. Attitude towards advertising
Attitude towards advertising refers to positive consumer inclination for specific advertisement (Lee and Cho 2011). In other words, attitude towards advertising manifests consumer's affinity and attitude towards specific advertisements. Attitude towards advertising has been studied by many scholars in marketing and advertising, since it was acknowledged as an important factor directly and indirectly affecting purchase intention (Sung and Cho 2011).

2.4.2. e-WOM; hashtag on SNS
Word-of-mouth (WOM) is defined as 'non-commercial activities of passing specific information about brand products or services by oral communication' (Engel, Kegerreis, and Blackwell 1969). The continual development of social media has prompted diverse research into online WOM intentions, called e-WOM. Consumers spreading e-WOM through SNS are motivated by opinion-seeking, opinion-giving, and opinion-passing

1

(Chu and Kim 2011). Opinion-seeking is a search behaviour for collecting information and advice required to make decisions. Opinion leaders use opinion-giving to impact consumers' attitudes and purchase intentions. Although the two types of e-WOM play significant roles in offline WOM, opinion-passing appears only on SNS, where a single click can allow consumers to forward and pass brand or product-related opinions or postings to large crowds. e-WOM in SNS is called 'share-of-post' (SOP) in which crowds learn about specific brands, products, and individual thoughts (Chatterjee 2011). Such e-WOM prompted the origin of hashtags originally used in the Internet Relay Chat (IRC) service to provide clustering benefits of consolidating information into a single position. However, e-WOM now is used mostly by young consumers to express specific support and interests. Accordingly, distinctive hashtag words and sentences are established as a new form of e-WOM that enables users to simultaneously share copious information while being entertained as they promote re-hashtag activities.

Corporations can expect positive e-WOM effects when they use SNS (Kim and Cho 2014) because consumers vigorously communicate and swiftly transmit messages regarding brands and products. Thus, corporate Facebook pages focus on e-WOM in accordance with the social distance of the source of information. e-WOM becomes the dependent variable to measure advertisement effects.

Nam, Son, and Lee (2015) identified that product, place, promotion, and price mediate the effects of SPA brands, and that SPA brands benefit most in offering reasonable prices. Also, price reduction, positive store reputation, and high distribution density positively affect brand equity (Yoo, Donthu, and Lee 2000). Lynch and Airely (1998) showed that price-oriented advertising, product information, and product experience directly impact purchasing. Thus, we hypothesize that SNS advertising of SPA brands affects attitudes toward advertising and e-WOM.

H4: There will be a difference in advertising effect (attitude towards advertising [H4-1], e-WOM [H4-2]) in accordance with the existence of the hashtag in SNS advertising of SPA brands and marketing activities.
H5: The perceived informativeness will positively (+) affect SNS advertising effectiveness of SPA brands (attitude towards advertising [H5-1], e-WOM [H5-2]).
H6: The perceived enjoyment will positively (+) affect SNS advertising effectiveness of SPA brands (attitude towards advertising [H6-1], e-WOM [H6-2]).
H7: Interactivity will positively (+) affect SNS advertising effectiveness of SPA brands (attitude towards advertising [H7-1], e-WOM [H7-2]).

Based on the literature, we intend to show that SNS advertising of SPA brands in accordance with hashtag and marketing activities generates positive perceived informativeness, enjoyment, interactivity, and advertising effects. Figure 1 shows the research model.

3. Research method

3.1. Research process

We conducted preliminary research to select appropriate SPA brand-related keywords stimuli based on preliminary research in which SPA brand-related keywords were derived from big data and in-depth interviews. To this end, we designed 4 [marketing activities (product/place/promotion/price)] × 2 [hashtag existence(Y/N)] variables. Through an

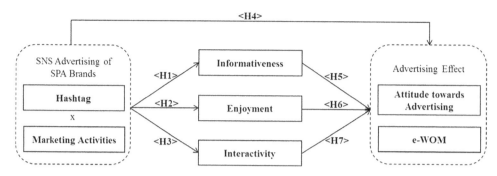

Figure 1. Research model.

online survey distribution and the convenience sampling method, we surveyed men and women in their 20s and 30s who use SNS. To prescreen survey participants, we asked: 'Are you familiar with SNS advertising?' Individuals who answered 'no' were thanked and removed from the respondent list. Of 1000 surveys, 890 were retrieved, but only 782 were used after incomplete or uncandid responses were removed. We administered eight survey variations, identified as A through H, so that 125 participants could be randomly categorized into eight groups receiving different treatments. All questions were identical except for the stimuli. We used SPSS 21.0 for frequency analysis, factor analysis, reliability analysis, two-way ANOVA, and regression analysis.

3.2. Measurement development

3.2.1. Preliminary test
Prior to measuring the advertising effect, this study deducted SPA brand-related keywords from real consumers in SNS through big data and in-depth interviews, owing to the fact that objective statistics or researches related to SNS advertisement of domestic SPA brands were lacking in numbers. According to preceding research, Twitter is able to collect the newest information and social issues due to their users distributing and sharing information from other media channels, thus making it a suitable platform for analyzing users' psychology (Ko et al. 2015). To extract data that illustrate SNS users' thoughts or feelings on SPA brands, social big-data research institute WizeNut's 'TweeTrend' was used for big-data analysis. Collected data were dated from 1 January 2015 to 24 September 2015 (approximately 9 months), and the keyword search method was utilized for data collection. This study selected and collected keywords from the top three global SPA brands that had entered the domestic market, sorted by sales figures. Details of keywords are as follows: 'SPA Brand', 'Fast Fashion', 'UNIQLO', 'H&M', and 'ZARA.' Every Korean and English keyword search result was provided in the form of an Excel file. Data filtering was completed for preconditioning, and data including articles, advertisements, negative comments, and foul words were excluded. Only consumer-written Twitter data were used in the analysis; a total of 2906 pieces of data were extracted. For text mining, morphological decomposition was executed using the 'Guljabi' program that most preceding researches made use of (Ko et al. 2015). A limitation in this study is that consumer-written social data are mostly unstructured text data, which are difficult to properly analyze and utilize, and that detailed analysis of public opinion on keywords is problematic (Park et al. 2011).

Table 1. Process of preliminary study.

	Big Data	In-depth interview	Final keywords
Product	pretty, good, a lot of, great, flimsy, beautiful, cute, etc.	various, trendy, simple, practical, comfortable, etc.	flimsy, light, cool, comfortable
Place	a lot of, large, convenient, easy, close, good, quick, etc.	convenient, big, spacious, exciting, accessible, etc.	spacious, accessible, light, convenient
Promotion	lack, good, cool, various, fine, different, new, etc.	interesting, must-have, new, special, etc.	limited period, new, special, exclusive
Price	good, cheap, fine, moderate, reasonable, etc.	reasonable, cheap, cost-effectiveness, etc.	cheap, reasonable, good, fine
Method of analysis	Morphological analysis and frequency analysis	Frequency analysis	Peer Review/goodness of fit Test

Therefore, this study conducted in-depth interviews to collect more diverse lexicons along with big data.

Fifteen interviews taking 40 minutes to an hour were conducted from 25–30 September 2015. Participants included eight fashion major students and seven fashion industry professionals. The interviewees were informed that the interviews were being recorded. Interviewers first briefly explained the purpose of the interview and then conducted a free association test to gather attitudes toward SPA brands. Then, participants answered questions about their marketing activities and freely described their opinions about product, place, promotion, and price in SPA brand marketing activities.

Data were extracted in accordance with high frequency and sorted by marketing activities. Table 1 shows the results of preliminary research of keyword extraction classified by SPA brand-marketing activities.

3.2.2. Stimuli selection

In this study, 4 [product/place/promotion/price] × 2 [hashtag/no hashtag] experimental design was used. The stimuli were made based on preliminary research involving SPA brand-related keywords from big data and in-depth interviews. The SNS platform in the experiment is Instagram, which has over 400 million monthly users as of September 2015, surpassing Twitter with 284 million monthly users (Park 2015). Instagram is an image content – predominant SNS – rather than text – on which hashtags are vigorously used (Moon, Lee, and Park 2015). In particular, regarding the rapid increase in fashion brands' use of Instagram to raise their brand awareness and promote their products (Ko et al. 2014), this research created advertisements modelled on the actual Instagram format.

This study intended to select SPA fashion brand subjects that were well known by experiment participants and were actively conducting marketing activities in SNS. According to Socialbakers (2015, 11), the only fashion brand that entered on the top-40 domestic brands page was SPA fashion brand UNIQLO Korea, with a total of 479,965 fans, becoming the number one domestic fashion brand page. Also, it has the highest market share with highest domestic sales (Koo 2015). In actuality, UNIQLO has seven fashion brand applications – being the most active SNS user among the SPA brands (Nam, Son, and Lee 2015). Accordingly, this study chose UNIQLO as the advertising stimulus, and the advertisements used as stimuli were real materials that UNIQLO had used previously in SNS. Three doctoral researchers majoring in fashion marketing and three fashion industry professionals, respectively, chose five advertisements of each marketing activities which had been promoted by UNIQLO within the last 6 months. The selected advertisements had their images extracted through prior investigation with 20 people specialized in fashion and clothing involved. It was completed after several modifications referring to results from preliminary research. The selected advertisement eliminated models, different typographies, and background images to minimize stimulating factors irrelevant to marketing activities. Other than keyword and images, every size and layout of advertisement was set the same to increase the internal validity and to prevent any differences caused by other controlling variables.

4. Results

4.1. Validity and reliability test

A factorial analysis of each measurement tool was done prior to hypothesis testing. Only principle component factor analysis and Varimax rotation were used as methods of factor extraction, and only those with an Eigen value of 1.0 or above and a factorial load of 0.5 or above were selected. Reliability analysis was conducted through reviewing internal consistency with the Cronbach's Alpha value, and was labelled as highly reliable when the CA value was 0.7 or higher, according to Nunnally (1978). The results of reliability test are as described in Table 2. In this research, the reliability of all factors on test values were higher than 0.7, achieving an internal consistency.

4.2. Hypotheses testing

H1, 2, 3, and 4 suppose that interactions between hashtag/no hashtag and marketing activities (product/place/promotion/price) will have discrepancies in affecting informativeness, enjoyment, interactions, and advertisement effect. The results from two-way ANOVA of these hypotheses are as described in Table 3.

The interaction effect on informativeness based on hashtags and marketing activities has been verified, and the result is shown in Figure 2. This result is identical to the in-depth analysis result of enterprises' Facebook posts and the advance research (Hwang

Table 2. Reliability test results.

Indicators	Questionnaires	Items	α
Enjoyment	This advertising is entertaining This advertising is pleasing This advertising is enjoyable	3	.941
Informativeness	This advertising is good sources of product/brand information This advertising is convenient source of information This advertising provides up-to-date information	3	.904
Interactivity	This advertising provides frequent exchange This advertising facilitates two-way communication This advertising can offer me a vivid communication experience	3	.873
e-WOM	I would recommend this ads to other users with using the hashtag I am willing to deliver(rehashtag) positive things of this ads I intend to share (like, hashtag, comments) this ads	3	.929
Attitude towards advertising	This advertising is favourable This advertising is appealing This advertising is interesting	3	.925

Note. α = Cronbach's α.

Table 3. The result of two-way ANOVA of hashtag and marketing activities.

Dependent variables	Independent variables	Mean square	F
Informativeness	Hashtag (A)	335.772	205.285***
	Marketing activities (B)	12.637	7.726***
	A × B	12.895	7.884***
Enjoyment	Hashtag (A)	146.639	89.531***
	Marketing activities (B)	0.096	0.059
	A × B	3.717	2.269
Interactivity	Hashtag (A)	203.435	97.403***
	Marketing activities (B)	7.668	3.672*
	A × B	1.723	0.825
Attitude towards advertising	Hashtag (A)	275.63	185.091***
	Marketing activities (B)	2.794	1.876
	A × B	6.904	4.636**
e-WOM	Hashtag (A)	168.733	94.957***
	Marketing activities (B)	11.247	6.33***
	A × B	9.14	5.144**

$^{*}p < .05, ^{**}p < .01, ^{***}p < .001.$

and Lim 2013) which shows advertisements on new products received the most reaction. Therefore, Hypothesis 1 was supported.

H2 was rejected since enjoyment does not keep interaction effects in mind, and H3 was rejected as it did not consider the interaction effect while suggesting hashtags and marketing activities would influence interactivity. This can be considered that the respondents did not recognize significant difference between enjoyment and interactivity on the hashtag marketing proposed unilaterally from companies. In other words, consumers would rather feel interested and recognize positive interactivity when exposed to bilateral marketing where they may communicate with others as well as participate, than unilateral marketing provided from companies (Chae, Shin, and Ko 2015; Ko et al. 2014).

H4, which suggests that hashtag and marketing activities can leverage advertisement effectivity, verified that attitude towards advertising and e-WOM had a statistically significant interaction effect. Therefore, both H4-1 and H4-2 were supported (Figures 3 and 4). According to Figure 4, the e-WOM of an advertisement that used hashtags in their price activity was higher than that of advertisements that exposed other activities to consumers. This result is identical to advanced research suggesting that reasonable price had the biggest effect among all of SPA's marketing activities (Kim and Koo 2014; Nam, Son, and

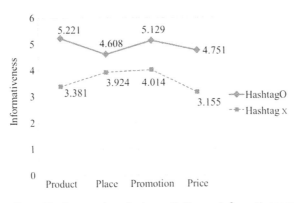

Figure 2. Interaction effect of hashtag and marketing activities on informativeness.

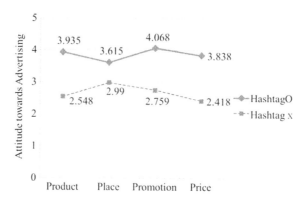

Figure 3. Interaction effect of hashtag and marketing activities on attitude towards advertising.

Lee 2015). Also, the main effect factors of hashtag activity (F = 94.957, $p < .001$) and marketing activities ($F = 6.329$, $p < .001$) were additionally identified.

H5, 6, and 7, which suggest that informativenes, enjoyment, and interactivity will have a positive effect on advertisement effect, were verified through multiple regression analyses, and the result is shown in Table 4. The result of H5, which is informativeness, and has a positive effect on attitude towards advertising, is as follows: the t-value for H5-1, informativeness affecting attitude towards advertising, was 13.946 ($p < .001$), and the t-value for H5-2, informativeness affecting e-WOM, was 10.605 ($p < .001$).

For H6, the result of multiple regression analysis of enjoyment has a positive effect on attitude towards advertising and is as follows: the value for enjoyment affecting attitude towards advertising was 20.294 ($p < .001$), and the same value for enjoyment affecting e-WOM was 15.839 ($p < .001$). Both factors seem to have an effect based on statistical level of significance, and H5 and 6 were therefore supported. This result is on the same path as advance research (Ducoffe 1996; Taylor, Lewin, and Strutton 2011) which showed that, informativeness and enjoyment have a positive effect on customers' attitudes towards advertising and usage intention.

The verification of H7, which suggests that the interactivity of SPA's SNS advertisement will positively affect advertisement effect, is as follows: the t-value for the H7-1, which includes interactivity having a positive effect on advertisement effect, was 17.286 ($p < .001$),

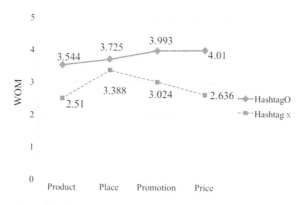

Figure 4. Interaction effect of hashtag and marketing activities on attitude towards e-WOM.

Table 4. The influence of informativeness, enjoyment, and interactivity on advertising effect.

Dependent variables	Independent variables	S.E	β	t-Value	R^2	F
Attitude towards advertising	(Constant)	0.094	–	4.291	.619	634.642***
	Informativeness	0.024	0.364	13.946		
	Enjoyment	0.026	0.530	20.294		
	Interactivity	0.027	0.526	17.286		
e-WOM	(Constant)	0.114	–	5.930	.494	380.379***
	Informativeness	0.030	0.320	10.605		
	Enjoyment	0.032	0.478	15.839		
	Interactivity	0.027	0.569	19.328		

***$p < .001$.

and was supported based on statistical level of significance. The *t*-value for H7-2, which includes interactivity having a positive effect on e-WOM, was 19.328 ($p < .001$), and was supported based on statistical level of significance. This result is identical to that of a previous research (Lee and Cho 2011) which suggested that higher interactivity will result in positive effect on advertisements.

5. Conclusion

5.1. Discussion

To counter the rapid turnover of merchandise and to maintain customer relationships, SPA brands conduct promotions and make special offers through various SNS marketing strategies (Chae, Shin, and Ko 2015). As a low-cost highly efficient advertisement, hashtags enable customers to share their participation and interest through e-WOM. However, companies would greatly benefit from objective and detailed research that identifies positive advertisement effects from hashtags. Accordingly, we undertook this research to discover how the marketing activities of SNS advertising of SPA brands affect informativeness, enjoyment, interactivity, attitudes toward advertising and e-WOM. We base our work on Ducoffe's (1996) web advertising model and on hashtags, one of the new e-WOM and marketing methods in SNS. Our study promotes the practical development of effective hashtag marketing strategies. Allure of hashtag, such as simplicity and entertainment, has been embraced by company as a marketing opportunity. It is one of the fastest ways to communicate with consumers and delivers effective e-WOM (Mcdowell 2017).

First, we used a two-way ANOVA to verify the interaction effect between hashtag and marketing activities and found the most positive statistically significant effect for informativeness of the advertisement. Thus, among the eight types of advertisements, informativeness was confirmed to be most important in web, mobile advertisements, as it is in traditional advertisement media (Brackett and Carr 2001; Ducoffe 1996). The result aligns with research showing that providing useful information is the main function of advertising, and that customers mainly seek useful information from advertisements. Additionally, we find that activity-promoting advertisements that use hashtags draw positive attitudes toward advertising, a result also aligned with previous research showing that consumers prefer emotional factors in product or brand advertising but prefer cognitive or rational factors in purpose-oriented advertisements such as sales or events (Strahilevitz and Myer 1998). e-WOM had the most positive effects among advertisements emphasizing lower prices. Indeed, SPA brands are successful mainly because they are inexpensive (Kim and

Yang 2015; Nam, Son, and Lee 2015). Also, we find that reasonable price activity positively affects brand attitudes.

Second, another *t*-test verified that hashtag/no hashtag yields different SPA advertisement effects in SNS. The results showed that hashtags evoked more positive reactions for any antecedents of SNS ads (informativeness, enjoyment, interactivity) and advertisement effects (attitudes toward advertising and e-WOM). In other words, customers evaluate every aspect of an advertisement more positively when businesses incorporate hashtags into their SNS advertisements. The finding supports research showing that tweets with hashtags enhance sharing activities (Lee and Jang 2013). Hashtags, which allow consumers to gather keywords of interest in one instance, are an excellent marketing method with benefits for both consumers and businesses. Furthermore, businesses can multiply their brand image or awareness through hashtags. Also hashtags are a new and powerful form of e-WOM allowing users to discuss certain topics, thus widening the scope of potential customers. Although brands may have high quality and renown, they must appeal to certain tastes if they are to evoke consumer interests. Accordingly, enterprises should understand the target consumers' interests through big data and prefer hashtag marketing as a way to advertise their products and improve their brand image.

Third, we verify that informativeness, enjoyment, and interactivity positively affect attitudes toward and effects of advertising. The result aligns with preceding research showing that informativeness and enjoyment significantly affect web advertising value and attitudes toward advertising (Ducoffe 1996), and that informativeness and enjoyment positively affect SNS advertisements (Taylor, Lewin, and Strutton 2011). Our results also align with research showing that interactivity positively affects attitudes toward advertising (Jee and Lee 2002; McMillan, Hwang, and Lee 2003). In short, consumers of SPA brands who find that SNS advertisements are informative, enjoyable, and interactive will have positive advertisement attitudes. Thus, informativeness, enjoyment, and interactivity are crucial.

5.2. Implications and limitations

This study is significant in that it focuses on the effect of hashtag utilization on SNS advertisements, a subject not previously treated in apparel study. Hashtags emerged fairly recently, so related research is inadequate. Nevertheless, hashtags greatly influence SNS: they are easy to use as a communication medium between brands and customers. Therefore, verification of the effect of hashtags on SNS fashion advertisements is a meaningful subject for study. Second, this research goes beyond traditional product or event-oriented advertisements to consider product, place, promotion, and price marketing activities of SPA brands as a new communication activity in the SNS advertisement area. Third, we verify that hashtag utilization has significantly positive effects on informativeness, enjoyment, interactivity, attitudes toward advertising, and e-WOM. As a result, fashion brands using hashtag methods may expect to encourage positive e-WOM if they understand the thoughts and interests of their customers. Fourth, we confirm that informativeness, enjoyment, and interactivity factors of SNS advertisements for SPA brands positively affect advertising results. In other words, companies should evoke customer interest and enjoyment by tying their interests to hashtags. We recommend that marketers create advertisements that encourage customer interactions with SNS and brands. Hashtags are a way to use common keywords of interest to form social consensus and emotional communications.

An indispensable factor in online marketing via SNS is the use of hashtags including simple keywords and users and location tags. The current generation is prone to communicate through online e-WOM that shares interests and opinions regarding brands. As a result, companies are using hashtags to reach customers and achieve superior marketing effects. Accordingly, in this research, we confirm the value of utilizing hashtags in SNS advertising for all marketers, not just SPA brands. Our research provides practical information for boosting the efficiency of hashtag marketing and e-WOM effects.

This study has some limitations that suggest avenues for follow-up studies. First, to exclude experimental artefacts, we used a real-life brand with a virtual stimulus as an advertisement. As a result, our ability to perfectly control factors such as brand preference was limited. Additionally, future research should use more meticulous criteria for selecting the stimuli. Second, we conducted this research analyzing big data through keyword reference as objectively and quantitatively as possible, but the linguistic nature of the data meant that we could not fully avoid subjectivity in selecting or interpreting keywords related to SPA brands. As a result, follow-up studies should develop a keyword selection method that excludes subjective views as much as possible.

Disclosure statement

No potential conflict of interest was reported by the authors.

References

Bezjian-Avery, A., B. Calder, and D. Iacobucci. 1998. New media interactive advertising vs. traditional advertising. *Journal of Advertising Research* 38, no. 4: 23–32.

Brackett, L.K., and B.N. Carr. 2001. Cyberspace advertising vs. other media: Consumer vs. mature student attitudes. *Journal of Advertising Research* 41, no. 5: 23–32.

Chae, H., E. Ko, and J. Han. 2015. How do customers' SNS participation activities impact on customer equity drivers and customer loyalty? Focus on the SNS services of a global SPA brand. *Journal of Global Scholars of Marketing Science* 25, no. 2: 122–41.

Chae, H., J. Shin, and E. Ko. 2015. The effects of usage motivation of Hashtag of fashion brands' image based SNS on customer social participation and brand equity: Focusing on moderating effect of SNS involvement. *Fashion & Textile Research Journal* 17, no. 6: 942–55.

Chatterjee, P. 2011. Drivers of new product recommending and referral behavior at social network sites. *International Journal of Advertising* 30, no. 1: 77–101.

Chaykowski, Kathleen. 2016. Instagram pushes into shopping with new mobile tools. *Forbes*, Nov 1. http://www.forbes.com.

Chu, S-C., and Y. Kim. 2011. Determinants of consumer engagement in electronic word-of-mouth (eWOM) in social networking sites. *International Journal of Advertising* 30, no. 1: 47–75.

Ducoffe, R.H. 1995. How consumers assess the value of advertising. *Journal of Current Issues and Research in Advertising* 17, no. 1: 1–18.

Ducoffe, R.H. 1996. Advertising value and advertising on the web. *Journal of Advertising Research* 36, no. 5: 21–35.

Duncan, T., and S.E. Moriarty. 1998. A communication-based marketing model for managing relationships. *Journal of Marketing* 62, no. 2: 1–13.

Engel, J.F., R.J. Kegerreis, and R.D. Blackwell. 1969. Word-of-mouth communication by the innovator. *Journal of Marketing* 33, no. 3: 15–19.

Hwang, J.S., and J.E. Lim. 2013. SNS as a strategic corporate communication tool: A content analysis of corporate Facebook fan-pages in Korea and the US. *The Korean Journal of Advertising* 24, no. 4: 143–78.

Jee, J., and W.N. Lee. 2002. Antecedents and consequences of perceived interactivity: An exploratory study. *Journal of Interactive Advertising* 3, no. 1: 34–45.

Jones, Sarah. 2014. Top 10 luxury brand social marketers of 2014. *Luxury Daily*, Dec. http://www.luxurydaily.com.

Kennedy, M.S., L.K. Ferrell, and D.T. LeClair. 2001. Consumers' trust of salesperson and manufacturer: An empirical study. *Journal of Business Research* 51, no. 1: 73–86.

Kim, Y.H. 2015. *Analysis and pattern of use of SNS (social network service)*. KISDISTAT report (15-03-02). Seoul: KISDI.

Kim, N.E., and J.J. Koo. 2014. An impact of SPA brand's marketing strategies on customers' purchase intention. *Korean Society of Basic Design & Art* 15, no. 5: 71–9.

Kim, H., and C. Cho. 2014. Effects of source's social distance on consumer's responses to corporate facebook page: Focusing on moderating effects of blatant persuasive intention, normative interpersonal influence and informative interpersonal influence. *The Korean Journal of Advertising* 25, no. 5: 7–42.

Kim, K.R., and S.J. Yang. 2015. Exploratory study on the success factors of SPA brands from marketing perspectives-based on grounded theory. *Journal of the Korean Society of Clothing and Textiles* 39, no. 2: 190–203.

Ko, E., E. Chun, S. Song, and P. Mattila. 2015. Exploring SNS as a consumer tool for retail therapy: Explicating semantic networks of "shopping makes me happy (unhappy)" as a new product development method. *Journal of Global Scholars of Marketing Science* 25, no. 1: 37–48.

Ko, J., J. Shin, E. Ko, and H. Chae. 2014. The effects of image contents based fashion brands' SNS toward flow and brand attitude: Focus on pleasure emotion as mediator. *Fashion & Textile Research Journal* 16, no. 6: 908–20.

Koo, H.R. 2015. Uniqlo, the first '1 trillion won fashion Brand'. *Joongang Ilbo*. http://news.joins.com.

Lee, S.Y., and C.H. Cho. 2011. Influences of interactive Ads' participation types on advertising effectiveness focusing on perceived interactivity of interactive film Ads. *The Korean Journal of Advertising and Public Relations* 13, no. 4: 95–124.

Lee, J.Y., and P.S. Jang. 2013. Effects of message polarity and type on word of mouth through SNS (social network service). *Journal of Digital Convergence* 11, no. 6: 129–35.

Lynch, J.G., and D. Ariely. 1998. Interactive home shopping: effects of search cost for price and quality information on consumer price sensitivity, satisfaction with merchandise, and retention. In *Marketing Science and the Internet, INFORM College on Marketing Mini-Conference*, March 6–8. Cambridge, MA.

Mcdowell, Meghan. 2017. In the age of Instagram, murals take on new meaning. https://www.businessoffashion.com.

McMillan, S.J., J.S. Hwang, and G. Lee. 2003. Effects of structural and perceptual factors on attitudes toward the website. *Journal of Advertising Research* 43, no. 04: 400–9.

Moon, H.N., Y.J. Lee, and S.H. Park. 2015. A study on change of gamification marketing through social media. Focusing on the marketing of Facebook fan page and Instargram hashtag. *Design Convergence Study* 14, no. 4: 209–21.

Nam, H.K., H.J. Son, and Y.R. Lee. 2015. Effect of SPA brand consumers' emotional consumption value orientation and assessment of marketing mix attributes on brand loyalty. *Journal of the Korean Society of Costume* 65, no. 4: 45–60.

Nunnally, J.C. 1978. *Psychometric theory*. New York: McGraw-Hill.

Oh, S.Y. 2015. Marketing, sulf in the sea of hashtag(#). *Marketing* 49, no. 10: 59–64.

Park, M.K. 2015. Instagram, 400 million users per month …40 billion shared photos. *Hankyung*. http://www.hankyung.com.

Park, K.I., and C.H. Cho. 2014. Factors influencing sharing activities in SNS: Focusing on moderating effects of social capital. *The Korean Journal of Advertising* 25, no. 5: 153–80.

Park, K., H. Park, H. Kim, and H. Ko. 2011. A study of opinion mining in SNS. *Journal of KIISE: Computer Systems and Theory* 29, no. 11: 54–60.

Shin, H.C. 2016. "Maximization of marketing effectiveness through SNS"… fashion market; fall in love with Hashtag. *Seoul Economic Daily*. http://news.naver.com.

Socialbakers. 2015. Socialbakers. http://socialbakers.com/statisticss

Sung, Y.H., and C.H. Cho. 2011. The effects of internet Ads on credibility and attitude toward the host site focusing on spillover effect. *The Korean Journal of Advertising and Public Relations* 13, no. 3: 448–81.

Strahilevitz, M.A., and G. Myer. 1998. Donation to charity as purchase incentives: How well they work may depend on what you are trying to sell. *Journal of Consumer Research* 24, no. 4: 434–46.

Tang, R., and P. Solomon. 1998. Toward an understanding of the dynamics of relevance judgment: An analysis of one person's search behavior. *Information Processing & Management* 34, no. 2: 237–56.

Taylor, D.G., J.E. Lewin, and D. Strutton. 2011. Friends, fans, and followers: Do Ads work on social networks? *Journal of Advertising Research* 51, no. 1: 258–75.

Yoo, B., N. Donthu, and S. Lee. 2000. An examination of selected marketing mix elements and brand equity. *Journal of the Academy of Marketing Science* 28, no. 2: 195–211.

Do we always adopt Facebook friends' eWOM postings? The role of social identity and threat

Yaeri Kim ⓘD, Yookyung Park ⓘD, Youseok Lee ⓘD and Kiwan Park ⓘD

ABSTRACT

In this research, we explore the role of social identity and threats to social identity on consumers' judgment and behavioural intention about electronic word-of-mouth (eWOM) on Facebook. Study 1 shows that sharing social identity with a Facebook friend increases perceptions of usefulness and behavioural intention to adopt eWOM. However, when a threat to social identity is posed, these positive effects are eliminated. Study 2 reveals an opposite condition wherein a threat to social identity results in associative responses to eWOM. When the social identity is perceived as impermeable (vs. permeable), threats that are posed toward the social identity increase perceived eWOM usefulness and adoption intention. eWOM source identification is revealed as an underlying mechanism explaining this relationship. Theoretical and managerial implications of these findings are discussed.

Introduction

Reading through the timeline on Facebook is an ordinary routine for Facebook users. Your Facebook 'friends' – who you know from going to school together or by having joined the same interest group on Facebook – are legitimate sources of information. For instance, you might find their posting of a review regarding their experience with a product interesting or helpful. However, can you really count on what your Facebook friends say? Whose posting would you consider to be more useful and, thus, be willing to adopt? The level of shared identity may matter. A friend with simply the same major at college may have less influence on electronic word-of-mouth (eWOM) adoption than a friend who has both the same major *and* belongs to the same student club. The nature of social identity may also be important. A friend based on impermeable (e.g. family members) versus temporary (e.g. acquaintance from a past workplace) relationships may also differ. Furthermore, what happens when the social identity based on the relationship between you and your friend faces social identity threat? In this paper, we aim to address these questions to gain a deeper understanding of consumers' judgments and behavioural intentions about eWOM on Facebook.

Interacting with one another on social networking sites (SNSs) is now a prevalent way of socializing and exchanging information. Remarkable progress in digital technology and wireless environments has allowed us to enjoy higher access to and more convenient sharing of information, accelerating deeper engagement in SNS communication. Social media usage is a global phenomenon, with almost half (47%) of smartphone owners visiting SNSs every day (Nielsen 2014). Facebook is the world's largest SNS, exceeding one billion active users, which means that one out of every seven people on the planet is on Facebook (Smith, Segall, and Cowley 2012). As of March 2016, 1.65 billion monthly active users were reported (King 2016).

Based on its collaborative and social characteristics, SNSs are a rising platform for consumer-to-consumer conversation, including eWOM, which is defined as 'any positive or negative statements made by potential, actual, or former customers about products or companies, which are made available to a multitude of people and institutions via the Internet' (Chu and Kim 2011; Hennig-Thurau et al. 2004, 39). However, what determines effective eWOM on SNSs has only been investigated in a few survey-based exploratory studies. For example, Teng et al. (2014) examined the role of message quality and source characteristics, and Chu and Kim (2011) incorporated social relationship factors, such as tie strength, homophily, trust, and interpersonal influence.

Extending this earlier research, we argue that the psychology of social identity (i.e. one's identity as defined by the social groups to which one belongs) (Tajfel and Turner 1986) is an important factor that determines consumers' eWOM adoption on Facebook. Facebook is currently the most popular SNS with such typical features (Boyd and Ellison 2007; Toma and Hancock 2013). More importantly, unlike conventional platforms (e.g. bulletin boards and online communities) where the source information is presented anonymously (Cheung et al. 2009; Dou et al. 2012), information regarding users' membership in social groups is explicitly public on Facebook. Social identity is exposed in consumers' profile, and consumers interact with each other based on common social groups, which often stem from real-life relationships (e.g. school, workplace). Moreover, we focus on a specific situation where social identity threat is posed. On SNSs, individuals' motives for positive self-expression and presentation are evident (Back et al. 2010; Gonzales and Hancock 2011; Wilcox and Stephen 2013). For instance, consumers edit and control what information is presented to others (Walther 2007) and selectively post desirable information about themselves (Gonzales and Hancock 2011). Thus, social identity threat may influence how individuals respond to eWOM due to their desire to maintain a positive self-image.

Drawing on the literature on eWOM and social identity (Tajfel and Turner 1986), this research provides the first empirical test of the possibility that social identity and threat to it may have an effect on the behavioural intention of eWOM adoption (i.e. liking, sharing, and commenting), as well as the assessment of eWOM message's informational value. Specifically, we first investigate whether perceptions of eWOM are impacted by the degree to which an individual shares their social identity with the source (study 1), as well as by the extent to which the social identity is perceived as impermeable (study 2). We also propose that social identity threat interacts with these characteristics, rendering distinct consequences.

Literature review and hypotheses

Social identity on Facebook

According to social identity theory, one's identity is comprised of personal identity (i.e. identity based on unique characteristics of the self) and social identity (i.e. identity related to group membership) (Tajfel and Turner 1986), which are both used to maintain positive self-worth (Steele 1988; Tajfel and Turner 1979; Tesser 2000). Along with the advent of SNSs, individuals are given opportunities to develop their social identity and build a complete sense of self by publicly displaying their social relationship and group membership (e.g. commenting on each other's walls and showcasing social connection) (Manago et al. 2008). Browsing one's Facebook profile, which includes information related to social identity, increases self-worth and self-integrity, and social connections to significant others provide self-affirmation (Sherman and Cohen 2006; Toma and Hancock 2013). Communicating on SNSs can even reduce feelings of loneliness and depression by providing an opportunity for self-disclosure and a platform from which to receive social support (Lee, Noh, and Koo 2013; Nimrod 2010). Indeed, it has been reported that Facebook users perceive themselves as having more social support than non-users (Hampton et al. 2011).

Unlike traditional online platforms, Facebook discloses users' profile information to the public, including acquaintances and strangers. Exposing one's profile information, including one's real name and affiliation to social groups, is highly encouraged on Facebook and is not regarded skeptical by users (Gross and Acquiti 2005). Such apparent openness is a distinct feature of Facebook compared to other SNSs that filter or discourage the disclosure of offline identity (e.g. dating websites) (Gross and Acquiti 2005). Moreover, Facebook provides various tools for displaying multiple aspects of the self (Young 2013). Most notably, aspects of social identity are made explicit by providing information on one's gender, age, and hometown. One's affiliation to specific social groups (e.g. educational institution, workplace) is also mentioned, and users can join 'Groups' within Facebook based on common hobbies, sports activities, and interests (Toma and Hancock 2013; Young 2013; Zhao, Grasmuck, and Martin 2008). In addition, the extent and depth of social relationships are portrayed by posting peer photographs and 'tagging' one another, which reveal that one is socially engaged and belongs to a certain group (Strano 2008; Zhao, Grasmuck, and Martin 2008). Furthermore, social identity information is utilized to identify one another and extend one's social network. Although many relationships on Facebook are 'anchored' in real-life social groups (Zhao 2006), someone who is not known offline or who is known indirectly through mutual friends may also be recommended as a 'friend' based on a common social group membership (Zhao, Grasmuck, and Martin 2008). For instance, following and 'liking' a sports team or alma mater allows users to discover and befriend other users sharing a similar social identity (Schmalz, Colistra, and Evans 2015).

Shared social identity and eWOM

On Facebook, 'friends' are the subjects of interactions and are considered sources of information through their activities, such as posting an opinion or uploading a link. However, Facebook friends differ from real-life friends. Whereas offline friends are able to convey varied meanings across relationships that differ in terms of intimacy and closeness, Facebook friends are based on simplified binary relations ('Friends or not'), which leads to less nuanced meanings. Hence, the threshold to be qualified as a friend is low on Facebook, which makes it difficult to specify the weight of the relationship, as they hold a different meaning than real-life friendships (Gross and Acquisti 2005). This leads to a fundamental question regarding the characteristics of Facebook friends that influence the evaluation of the eWOM message.

In the current research, we propose that the sharing of social identity between the receiver and the source (Facebook friend) may influence one's perception of the informational value of the eWOM message because social identity information is salient and important on Facebook. The shared level of social identity with the source refers to how much one's social identity overlaps with that of the source, which is based on how much the two individuals belong to the same social group. It is related to awareness of joint membership within a social group (Tajfel 1978) and results from categorizing oneself and the source as having a collective social identity (Dholakia et al. 2009). Such cognitive process may include judgments regarding how similar (dissimilar) one is to another who shares (does not share) a social identity (Mousavi, Roper, and Keeling 2017).

Social identity sharing may have a profound effect on the manner in which the source and eWOM are evaluated for several reasons. First, belonging to the same social group creates a sense of conformity (Cialdini and Goldstein 2004). Group norms are developed and enforced to regulate in-group members' behaviour (Feldman 1984), and members ought to conform to the group's point of view due to both normative and informative influences (Eagly and Chaiken 1993). This may lead to higher susceptibility to eWOM from a source who shares greater social identity. Second, social group members tend to share common attributes (e.g. gender, age, race, and education), which signify homophily (Rogers and Bhowmik 1970). Importantly, interpersonal communication is more likely to occur among homophilous members (Lazarsfeld and Merton 1954; Rogers 1995; Rogers and Bhowmik 1970), and homophilous sources are more likely to be used as information sources (Brown and Reingen 1987). Homophily also leads to a greater level of interpersonal attraction and trust (Ruef, Aldrich, and Carter 2003), and it increases persuasion (Chu and Kim 2011; Walther, Slovacek, and Tidwell 2001; Wang et al. 2008). In particular, similarity between the source and the receiver increases the persuasive influence of WOM and eWOM (e.g. Feick and Higie 1992; Kiecker and Cowles 2002; Reichelt, Sievert, and Jacob 2014) because people want to interact with those who are similar to themselves (Laumann 1966; Brown and Reingen 1987). Thus,

H1: Social identity sharing between the source and the receiver will increase (H1a) perceptions of eWOM usefulness and (H1b) behavioural intention about eWOM adoption.

In addition, certain characteristics of social identity that are shared by the receiver and the source might also influence receivers' tendency toward eWOM adoption. In this research, we attend to the fact that some social groups are perceived as relatively hard to change, while others may be easily joined and withdrawn from. This refers to the perception of (im)permeability of a group, which is the extent to which individual group members can(not) leave one group and join another (Verkuyten 2005). If a group boundary is impermeable, then withdrawing membership is regarded as almost impossible, as in the case of age and race. Similarly, Jackson and colleagues (1996) have discussed that groups can be perceived as impermeable if group membership is stable across the lifespan or has permanent characteristics. In other words, the concept of social identity's impermeability lies in consumers' perception. For instance, one's affiliation to a company or residential area may be regarded as permanent or temporary based on how the member feels bounded by the membership.

Findings from past studies provide some clues that help predict consumers' attitudes toward eWOM based on the extent of perceived group impermeability. For instance, when group boundaries are perceived as impermeable, people identify more with the group and are more likely to define themselves as a group member rather than at the individual level (Ellemers, Spears, and Doosje 2002; Tajfel 1975, 1978). In addition, there is a tendency to be less competitive and exhibit greater personal sacrifice for the group when the group is impermeable (Ellemers, Wilke, and Van Knippenberg 1993) because individuals are reluctant to act in a selfish manner when the group boundary is perceived as impermeable. Therefore, in the context of Facebook, we posit that perceived impermeability of social identity will positively influence the perception of eWOM messages. Accordingly, consumers will evaluate eWOM more favourably when they share an impermeable group membership with the source. Hence,

H2: Perceived impermeability of shared social identity between the source and the receiver will increase (H2a) perceptions of eWOM usefulness, and (H2b) behavioural intention about eWOM adoption.

Social identity threat and eWOM

When social identity is temporarily devalued or marginalized (e.g. receiving negative information about gender identity), it results in a threatened social identity (Major and O'Brien 2005). In these situations, individuals predominantly attempt to cope with the threat and recover positive social identity (Tajfel and Turner 1979). In this paper, we attend to the social mobility strategy which refers to an individual-level approach of leaving or dissociating themselves from the group that shares the same social identities (Tajfel and Turner 1979; Jackson et al. 1996). Individuals deliberately attempt to decrease the perception that they belong to the threatened group (Ellemers, Spears, and Doosje 1997) and emphasize heterogeneity within the threatened group (Doosje et al. 1999) whose negative group features may be transmitted to the individual self. Individuals may psychologically depart from the threat by distancing themselves from the group. For instance, they may reluctantly identify with the group, reduce perceived similarity to the group, or decrease interacting with group members. Moreover, individuals may actually relinquish group membership if such an attempt is viable (Jackson et al. 1996). Thus, when a threat is presented, any positive attitudes toward highly homophilous sources who share the same social identity will disappear.

In the marketing field, consumers respond by preferring identity-unrelated products and avoiding threatened identity-related products as a means of defending and upholding their feelings of self-worth (Tepper 1994; White and Argo 2009; White and Dahl 2007). In the context of SNSs, we propose that threats to a shared social identity will lead to an individual's dissociation from the group that shares the particular threatened social identity. As such, individuals under identity threat are expected to denigrate the credibility of the in-group source, thereby decreasing subsequent perceptions of eWOM's informative value and behavioural intention about eWOM adoption. Thus,

H3: The positive effect of social identity sharing on (H3a) perceptions of eWOM usefulness, and (H3b) behavioural intention about eWOM adoption will disappear when there is a threat posed to social identity.

When facing a shared social identity threat, the first approach (i.e. social mobility strategy) involves using a dissociative strategy to psychologically depart from the threatened social identity, as has been described in the previous section. However, several studies have introduced another coping strategy as an alternative to avoiding or dissociating oneself from the in-group that shares the same social identities. Ellemers, Spears, and Doosje (1997) proposed that responses to group identity threat can differ depending on the individual's prior level of in-group identification. Low identifiers are likely to implement dissociative strategies by departing from the in-group, whereas high identifiers are likely to respond with stronger group cohesion. White and Argo (2009) also posit that the level of collective self-esteem (CSE) affects individual coping strategies in the face of social identity threat. Those who are low in CSE do not identify with or value the social group, and they may take a self-protective response by dissociating from the group. On the other hand, those who are high in CSE identify with and value the social identity, and they may maintain their group associations in the face of threat. Furthermore, White, Argo, and Sengupta (2012) discovered that when those with more independent self-construals experience a social identity threat, they avoid threatened identity-linked products to restore their individual self-worth. In contrast, those with highly interdependent self-construals use associative strategies (i.e. a greater preference for identity-linked products) to satisfy their need to belong. Therefore, depending upon the level of in-group identification, CSE, and self-construal, individuals may adopt associative responses when social identities are threatened.

Consistent with social identity threat and perspectives regarding associated coping strategies, we propose (im)permeability as a new factor that may influence whether or not individuals adopt an alternative coping strategy. In particular, when social identity is threatened, one of the distinguishing socio-structural factors that determines whether an individual uses social mobility strategies is the perceived impermeability of group boundaries (Verkuyten and Reijerse 2008). For example, Ellemers, Wilke, and Van Knippenberg (1993) have shown that individuals are reluctant to use social mobility strategies (i.e. dissociating from the group) when the social identity is perceived as impermeable. This is consistent with prior studies that people show stronger in-group identification when the social group boundary is perceived as impermeable, even when the in-group is viewed negatively (Ellemers et al. 1988; Ellemers, Knippenberg, and Wilke 1990).

Strongly identifying with a certain social identity is a sign of solidarity and commitment toward that social identity since renouncing it would signify that one has lost an important part of the self (Schmalz, Colistra, and Evans 2015). When this commitment is high, the homogeneity of the in-group is stressed even under a threat (Doosje, Ellemers, and Spears 1995). That is, members' intention to exit and join another group does not increase, even when they are presented with negative information regarding the impermeable group (Lalonde and Silverman 1994). Rather, identification with the in-group increases in such cases. However, if the commitment toward the group is low, exposure to negative group information will lead one to adopt social mobility strategies (Doosje et al. 1999). Thus, we expect that social identity threat may increase individuals' association with the social identity when it is perceived as impermeable. In the context of SNSs, this associative

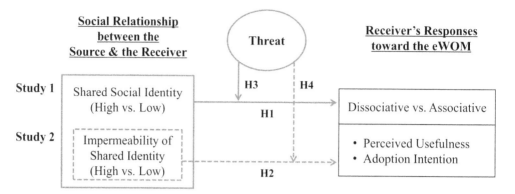

Figure 1. Overview of the research.

response would be represented by supportive actions toward the eWOM source who shares the threatened social identity and by increased eWOM usefulness and adoption. Hence,

H4: The positive effect of perceived impermeability of the social identity on (H4a) perceptions of eWOM usefulness and (H4b) behavioural intention about eWOM adoption will be more pronounced under conditions of social identity threat (see Figure 1 for the research overview).

Study 1

The objective of study 1 was to test the hypothesis that the shared level of social identity between the source and the receiver would positively influence the judgment of eWOM on Facebook. A total of 387 participants (198 females, 189 males) who had a Facebook account were recruited via Amazon's Mechanical Turk, and randomly assigned to one of four conditions based on a 2 (shared social identity: high vs. low) × 2 (social identity threat: threat vs. no threat) between-participants design.

Pre-test

Prior to the main study, a pre-test was conducted to confirm in advance the manipulation scenario of shared social identity. A total of 132 participants (75 females, 57 males) were randomly presented with one of two conditions of shared social identity: high vs. low. First, social identity in terms of high school membership was made salient. To do so, participants were asked to think about their membership of high school alumni (Luhtanen and Crocker 1992). They wrote down the name and location of the high school they went to or graduated. Then, participants were instructed to imagine themselves browsing Facebook and spotting a friend named 'Pat'. In the *high sharing* condition, Pat was described as a high school alumnus while Pat was not a high school alumnus in the *low sharing* condition. Next, participants answered how much Pat shares social identity based on two measurements. First, a set of seven-step Venn diagrams were presented. Each diagram represented different degrees of overlap of two circles that indicate the participant and

Pat. If the circles' overlapped area is larger, this reflected greater shared social identity between the participant and Pat (adapted from Aron, Aron, and Smollan 1992). As expected, participants in the high sharing condition felt that Pat shared greater level of social identity (M_{high} = 2.484, M_{low} = 1.956; $F(1, 130)$ = 5.520, $p < .05$). Second, participants directly answered the degree of shared social identity they perceived of Pat upon a percentage scale (0% = none shared; 100% = very much shared). Again, participants in the high sharing condition indicated a greater level of shared social identity with 'Pat' (M_{high} = 32.734, M_{low} = 25.956; $F(1, 130)$ = 2.993, $p = .086$) albeit marginally significant.

Method

All participants were asked to recall their membership of high school alumni and write down the name and location of their high school as in the pre-test. Then, participants took part in two unrelated studies. The first part involved manipulating social identity threat, in which participants read about intellectual achievements of their high school alumni over the last 10 years, ostensibly retrieved from National Center of Education Statistics (adapted from Dietz-Uhler and Murrell 1999). In the *threat* condition, the level of analytical reasoning skill, motivation in the workplace, and sense of social intelligence were described as being evaluated poorer than the national average. In the *no-threat* condition, the achievements were similar to the national average. In the second part, participants were instructed to imagine themselves browsing Facebook and reading a review post from a Facebook friend named Pat. The descriptions of Pat varied using the manipulation in pre-test for the level of shared social identity (high vs. low). The posting consisted of Pat's comments upon visit to a restaurant along with four photos (see Appendix).

Subsequently, participants completed a questionnaire that includes measures pertaining to the evaluations of Pat's posting, expertise on restaurants, and demographics. Adapted from Sussman and Siegal (2003), eWOM usefulness was measured using 7-point semantic differential items, anchored with 'worthless/valuable, uninformative/informative, and harmful/helpful' (α = .837). eWOM adoption intention was measured by assessing the willingness to engage in three typical behaviours in Facebook: pressing 'Like' on a posting, leaving a 'Comment' on a posting, and 'Sharing' a posting. The three items were measured on a 7-point scale (1 = very low; 7 = very high, α = .837). As a control variable, expertise on restaurant was measured with three items (e.g. 'I am familiar with restaurants in general,' 'I have much experience with restaurants in general') on a 7-point scale (Ryu and Han 2009) (1 = strongly disagree; 7 = strongly agree, α = .927).

Results and discussion

As a check on social identity threat, participants indicated their agreement with each of the following statements on a 7-point scale (1 = strongly disagree; 7 = strongly agree, α = .816): 'I feel uncomfortable when I receive this information about my high school,' 'I feel threatened when I read this information about my high school,' 'My feelings are negative when I read the information about my high school' (Jetten, Postmes, and McAuliffe 2002). Participants in the threat condition felt more threatened than those in the no-threat condition (M_{threat} = 3.903, $M_{no-threat}$ = 2.713; $F(1, 382)$ = 59.187, $p < .000$).

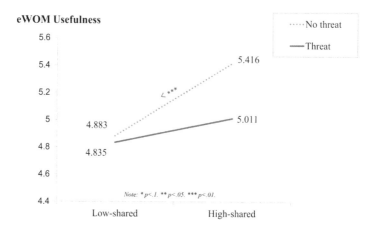

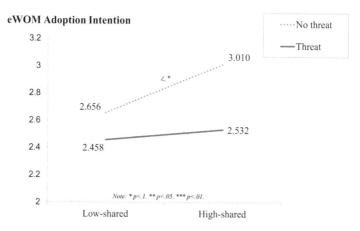

Figure 2. eWOM usefulness and eWOM adoption intention as a function of level of shared social identity and social identity threat (study 1).

Each of the two dependent measures, eWOM usefulness and eWOM adoption intention, was entered to a 2 (shared social identity) × 2 (social identity threat) analysis of covariance (ANCOVA) with restaurant expertise included as a covariate (see Figure 2). First, we analysed the effects of shared social identity and its threat on eWOM usefulness. A significant main effect of shared social identity on eWOM usefulness was found (M_{high} = 5.216, M_{low} = 4.859; $F(1, 382)$ = 10.411, p = .001). The main effect of social identity threat was also significant (M_{threat} = 4.923, $M_{no-threat}$ = 5.150; $F(1, 382)$ = 3.912, p = .049). Most importantly, the two-way interaction on eWOM usefulness revealed significant as predicted ($F(1, 382)$ = 3.980, p = .047). Contrast analyses provided support for H1a that participants in the high sharing condition perceived eWOM from Pat as more useful than those in the low sharing condition (M_{high} = 5.416, M_{low} = 4.883; $F(1, 382)$ = 13.765, p < .000) when there is no threat. However, when threat was posed to the social identity, the positive effect of shared social identity was eliminated (M_{high} = 5.011, M_{low} = 4.835; $F(1, 382)$ = .778, p = .378) supporting H3a.

Second, influences on eWOM adoption intention were examined. The main effect of shared social identity on eWOM adoption intention was insignificant (M_{high} = 2.775,

$M_{low} = 2.556$; $F(1, 382) = 2.492$, $p = .115$) and the main effect of social identity threat was significant ($M_{threat} = 2.495$, $M_{no\text{-}threat} = 2.833$; $F(1, 382) = 5.592$, $p = .019$). The two-way inter-action on eWOM adoption intention was insignificant ($F(1, 382) = .746$, $p = .388$). However, contrast analyses indicated a similar pattern to what was found for eWOM usefulness. Participants in the high sharing condition showed higher adoption intention for eWOM ($M = 3.010$) than those in the low sharing condition ($M = 2.656$; $F(1, 382) = 3.013$, $p = .083$) when the threat was not posed (H1b), albeit marginally significant. Yet, when participants were exposed to threat (H3b), the influence of shared social identity diminished ($M_{high} = 2.532$, $M_{low} = 2.458$; $F(1, 382) = .261$, $p = .610$).

The results from this study demonstrate that consumers appreciate more of the posting in Facebook and are more willing to 'like', 'comment', and 'share' it when the source (i.e. Facebook friend) shares social identity with them. Nevertheless, such positive impacts disappeared when their social identity was threated. This is similar to prior studies that revealed dissociative responses upon identity threat (e.g. White and Argo 2009).

Study 2

The objective of study 2 was to examine whether the perceived impermeability of a social identity may induce distinguished consequences of reinforcing the positive attitude toward eWOM even upon threat. We also aimed to demonstrate that the source identification exists as an underlying mechanism. A total of 196 responses (105 females, 91 males) from Amazon's Mechanical Turk were analysed using a 2 (perceived impermeability of social identity: high vs. low) × 2 (social identity threat: threat vs. no threat) between-participants design. This study postulates that the source shares social identity with the participant and only differs in terms of how impermeable the shared social identity is. Also, we utilized a different social identity of being a member of a company's membership program to generalize our findings.

Method

Participants were presented with two parts as in study 1. In the first part, participants read a scenario in which they were a member of 'ETNA,' a fictitious American membership-only supermarket chain. In the *high impermeable* condition, participants were imagined to have lifetime membership with no expiration date. The registration fee of $100 was described as non-refundable. In the *low impermeable* condition, participants were imagined to have two-year membership which would expire this month. The registration fee of $100 was described as refundable according to remaining period. Next, participants read a news article with study results on green consumption of US citizens to manipulate social identity threat. The fictitious investigation covered consumers from the 50 national membership-only supermarket chains and stated how the members of ETNA were evaluated. In the *threat* condition, ETNA members were evaluated as poorer than the national average on dimensions such as the amount in-store garbage, excessive usage of plastic bags, environmentally friendly product purchase, and recycling participation. Participants were also told that their regional community responded sensitively to the results and criticized ETNA members. In *no-threat* condition, ETNA performance for these criteria was similar to the national average, and their regional community was interested with the results. In the second part, participants imagined themselves browsing Facebook and reading a posting of a Facebook friend *Pat* who was also a member of ETNA and shared Facebook friends who were ETNA members as well. The posting was identical with the stimulus used in study 1.

Participants completed a questionnaire that included measures of manipulation checks, source identification, perceived eWOM usefulness, behavioural intention about eWOM adoption, and demographics. Adapted from Mummendey and Wenzel (1999) and Dholakia, Bagozzi, and Pearo (2004), perceived source identification was measured on a 7-point scale (1 = strongly disagree; 7 = strongly agree, α = .943) by asking how much they agreed with the four statements: 'I see myself and Pat as a member of a same social group,' 'Pat and I can be categorized as members of same social group,' 'Pat and I belong to a same social group,' 'Pat and I have same social identity'. Perceived eWOM usefulness (α = .809) and behavioural intention about eWOM adoption (α = .865) were measured using the same items as in study 2.

Results and discussion

To verify the manipulation of perceived impermeability, participants responded to the following statement about 'ETNA' membership: 'In your opinion, how difficult is it for you to withdraw ETNA membership? (1 = not very difficult; 7 = very difficult), adapted from Jackson et al. (1996).' As expected, participants in the high condition felt that the ETNA membership is more impermeable than those in the low condition (M_{high} = 4.480, M_{low} = 3.020; $F(1, 192)$ = 28.376, p < .000). Manipulation for social identity threat was checked with the same items used in study 1 (α = .875). As intended, a higher level of threat was produced for participants in the threat condition (M_{threat} = 4.709, $M_{no-threat}$ = 2.941; $F(1, 192)$ = 74.785, p < .000).

We analysed eWOM usefulness and eWOM adoption intention using a 2 (perceived impermeability) × 2 (social identity threat) ANOVA (see Figure 3). The main effect of perceived impermeability of social identity (M_{high} = 5.211, M_{low} = 5.000; $F(1, 192)$ = 2.399, p = .123) and the main effect of social identity threat (M_{threat} = 5.124, $M_{no-threat}$ = 5.088; $F(1, 192)$ = .062, p = .804) on eWOM usefulness was insignificant. Yet, a significant interaction of perceived impermeability and its threat ($F(1, 192)$ = 4.958, p = .027) was found. Contrast analyses showed that participants high and low in impermeability perceived similar level of usefulness (M_{high} = 5.039, M_{low} = 5.137; $t(192)$ = −.489, p = .625) when there was no threat, not supporting H2a. However, when participants were exposed to threat, higher impermeability led to higher perception of eWOM usefulness (M_{high} = 5.397, M_{low} = 4.851; $t(192)$ = −2.617, p = .010) supporting H4a.

The analysis on eWOM adoption intention showed insignificant main effect of perceived impermeability (M_{high} = 2.990, M_{low} = 2.684; $F(1, 192)$ = 2.022, p = .157) and social identity threat (M_{threat} = 2.940, $M_{no-threat}$ = 2.742; $F(1, 192)$ = .750, p = .388). The interaction revealed significant ($F(1, 192)$ = 4.090, p = .045) consistent to the results for eWOM usefulness. Contrast analyses showed that there was no significant difference between participants high and low in impermeability (M_{high} = 2.673, M_{low} = 2.811; $t(192)$ = −.434, p = .665) rejecting H2b. However, when threat was posed, higher impermeability of social identity led to higher willingness to adopt eWOM (M_{high} = 3.333, M_{low} = 2.546; $t(192)$ = −2.387, p = .018) supporting H4b.

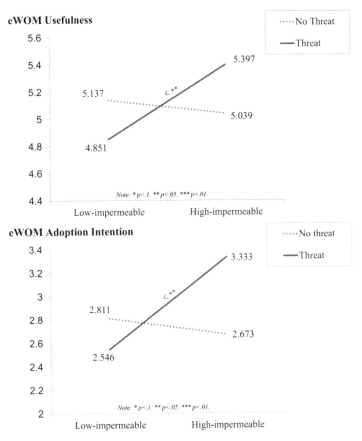

Figure 3. eWOM usefulness and eWOM adoption intention as a function of perceived impermeability of social identity and social identity threat (study 2).

In addition, a moderated mediation analysis tested the mediating role of source identification to explain the effect of perceived impermeability on eWOM usefulness and eWOM adoption (see Figure 4). The bootstrapping technique for conditional indirect effects (Preacher and Hayes 2008) estimated a significant indirect effect among participants in threat condition: perceived impermeability increased eWOM usefulness (95% CI = [.031, .336]) and eWOM behavioural intention (95% CI = [.058, .582]) through increased perception of source identification. Yet, there was no corresponding indirect effect in no-threat condition on both dependent variables: perceived impermeability had no effect on eWOM usefulness (95% CI = [−.330, .012]) and eWOM adoption intention (95% CI = [−.575, .031]) through the mediator.

Results from this study revealed that consumers evaluate the posting of a Facebook friend who shares an impermeable social identity as more favourable even and only when the identity was threatened. Moreover, the underlying mechanism was found that perceived impermeability of social identity increased the respondents' perception that they belonged to the same group as the source which resulted in amplified associative responses toward the source's eWOM when social identity threat was posed. For those with temporal thus less impermeable social identity, receiving social identity threat did

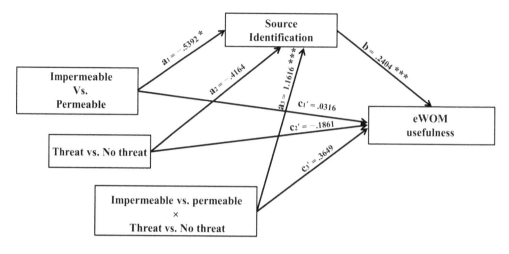

Note: * p<.1. ** p<.05. *** p<.01.
Coding: Impermeability (1), Permeability (0); Thereat (1). No threat (0)

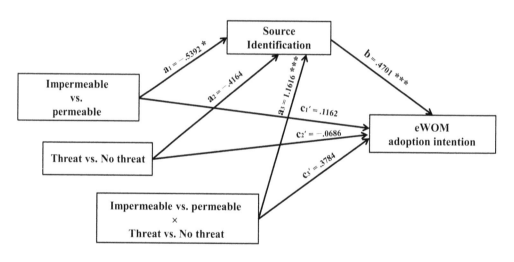

Note: * p<.1. ** p<.05. *** p<.01.
Coding: Impermeability (1), Permeability (0); Thereat (1). No threat (0)

Figure 4. Mediating role of source identification (study 2).

not boost the desire to identify with the group because they could avoid the criticism easily.

General discussion

The current research examines how individuals' responses to eWOM on Facebook differ depending on whether or not the individuals' social identity is highly shared with that of the source (study 1) and whether social identity is perceived as impermeable (study 2). In study 1, participants assessed eWOM more positively when they shared a high school identity with the source. However, this effect was minimized when the identity was undervalued. In study 2, participants in the impermeable condition rated eWOM useful-ness and adoption intention higher even though the identity was criticized. In addition, source identification emerged as the underlying mechanism accounting for this causal relationship.

This research contributes to the literature on eWOM and social identity in several ways. First, to the best of our knowledge, the current research is the first to apply social identity theory to eWOM in the context of Facebook. Even though a few studies have explored social relationship factors that contribute to eWOM in SNSs context (e.g. Chu and Kim 2011), no studies have examined the role of social identity and threats to it. Second, in contrast to prior research that has examined eWOM in SNS contexts through surveys, this research used an experimental paradigm to elucidate causal relationships among varia-bles. Third, this research utilized real (i.e. high school) and fictitious social identity (i.e. supermarket membership) conditions to generalize findings. Finally, the present study introduces a realistic and context-specific variable, labelled behavioural intention about eWOM adoption, which was measured by asking respondents' intention to 'Like' a post-ing, leave a 'Comment' on it, and 'Share' it. This newly constructed variable captures typi-cal reactions of Facebook users in reality so that future research can effectively adopt this construct to explore SNS behaviours.

Although this study provides initial findings on eWOM, a few limitations should be noted. First, although the present research covers major aspects of social identity theory (e.g. identity sharing, identity threat, and impermeability of identity), other aspects of social identity need to be considered in explaining eWOM-related judgment and behav-iour. For instance, the length of time after acquiring social identities, differences between an innate identity and an acquired identity, and the level of effort invested to obtain iden-tities are fruitful alternatives for future research. Second, not only can the existence of identity threat moderate the effect of social identity, but the type of threat can do so as well. Han, Duhachek, and Rucker (2015) argue that individuals respond differently depending on which aspect of identity is threatened. Third, the level of identity sharing and impermeability may have an additional effect on active behaviours, such as generat-ing one's own eWOM messages. Finally, although we had initially expected that the per-ceived impermeability of social identity would positively affect eWOM evaluation, it was restricted to situations in which social identity was threatened. A potential explanation for this is because the fictitious supermarket membership utilized in the scenario may not have been considered as realistic or perceived as important social identity to the partici-pants. If the group membership reflected more realistic factors in life and was regarded as more critical (e.g. member of workplace or village), then perceived impermeability of the social identity may work as a significant factor on eWOM evaluation even under no-threat condition. In other words, the positive effects of perceived impermeability on eWOM eval-uation might be restrictively activated depending on the significance of social identity or related situations. Thus, we contend that when and why perceived impermeability per-forms as a crucial factor on eWOM evaluation may also be an important area to explore in the future research.

Findings from this research yield various managerial insights for eWOM management and membership program management. First, marketers should consider social identity characteristics of target populations when exploiting network hubs or opinion leaders to spread eWOM messages. Specifically, transmitters who share wholesome social identities with target customers may exert stronger influence on receivers. However, when a social threat to shared identity is unavoidable, marketers need to effectively strategize in order to make receivers react associatively to the source. Second, companies' membership programs can serve as a new form of social identity in capitalistic economies. As a result, companies must strengthen the group network among consumers and leverage the network as a mouthpiece to diffuse favourable messages not only to group members, but across the entire market as well. Furthermore, since the current research confirms the positive impact of identity's impermeability, companies should encourage customers to maintain membership for an extended period of time or effectively lock them in the program.

Disclosure statement

No potential conflict of interest was reported by the authors.

Funding

This research was financially supported with a grant from the Institute of Management Research, Seoul National University.

ORCID

Yaeri Kim ⓘ http://orcid.org/0000-0002-4311-2075
Yookyung Park ⓘ http://orcid.org/0000-0003-4729-1223
Youseok Lee ⓘ http://orcid.org/0000-0002-2882-9481
Kiwan Park ⓘ http://orcid.org/0000-0001-8320-9560

References

Aron, A., E.N. Aron, and D. Smollan. 1992. Inclusion of other in the self scale and the structure of inter-personal closeness. *Journal of Personality and Social Psychology* 63, no. 4: 596–612.

Back, M.D., J.M. Stopfer, S. Vazire, S. Gaddis, S.C. Schmukle, B. Egloff, and S.D. Gosling. 2010. Facebook profiles reflect actual personality, not self-idealization. *Psychological Science* 21, no. 3: 372–74.

Boyd, D.M., and N.B. Ellison. 2007. Social network sites: Definition, history, and scholarship. *Journal of Computer—Mediated Communication* 13, no. 1: 210–30.

Brown, J.J., and P.H. Reingen. 1987. Social ties and word-of-mouth referral behavior. *Journal of Consumer Research* 14, no. 3: 350–62.

Cheung, M.Y., C. Luo, C.L. Sia, and H. Chen. 2009. Credibility of electronic word-of-mouth: Informational and normative determinants of on-line consumer recommendations. *International Journal of Electronic Commerce* 13, no. 4: 9–38.

Chu, S.C., and Y. Kim. 2011. Determinants of consumer engagement in electronic word-of-mouth (eWOM) in social networking sites. *International Journal of Advertising* 30, no. 1: 47–75.

Cialdini, R.B., and N.J. Goldstein. 2004. Social influence: Compliance and conformity. *Annual Review of Psychology* 55: 591–621.

Dholakia, U.M., R.P. Bagozzi, and L.K. Pearo. 2004. A social influence model of consumer participation in network-and small-group-based virtual communities. *International Journal of Research in Marketing* 21, no. 3: 241–63.

Dholakia, U.M., V. Blazevic, C. Wiertz, and R. Algesheimer. 2009. Communal service delivery: How customers benefit from participation in firm-hosted virtual P3 communities. *Journal of Service Research* 12, no. 2: 208–26.

Dietz-Uhler, B., and A. Murrell. 1999. Examining fan reactions to game outcomes: A longitudinal study of social identity. *Journal of Sport Behavior* 22, no. 1: 15–27.

Doosje, B., N. Ellemers, and R. Spears. 1995. Perceived intragroup variability as a function of group status and identification. *Journal of Experimental Social Psychology* 31, no. 5: 410–36.

Doosje, B., R. Spears, N. Ellemers, and W. Koomen. 1999. Perceived group variability in intergroup relations: The distinctive role of social identity. *European Review of Social Psychology* 10, no. 1: 41–74.

Dou, X., J.A. Walden, S. Lee, and J.Y. Lee. 2012. Does source matter? Examining source effects in online product reviews. *Computers in Human Behavior* 28, no. 5: 1555–63.

Eagly, A.H., and S. Chaiken. 1993. *The psychology of attitudes*. Orlando, FL: Harcourt Brace Jovanovich College Publishers.

Ellemers, N., A.V. Knippenberg, N.D. Vries, and H. Wilke. 1988. Social identification and permeability of group boundaries. *European Journal of Social Psychology* 18, no. 6: 497–513.

Ellemers, N., A. Knippenberg, and H. Wilke. 1990. The influence of permeability of group boundaries and stability of group status on strategies of individual mobility and social change. *British Journal of Social Psychology* 29, no. 3: 233–46.

Ellemers, N., R. Spears, and B. Doosje. 1997. Sticking together or falling apart: In-group identification as a psychological determinant of group commitment versus individual mobility. *Journal of Personality and Social Psychology* 72, no. 3: 617–26.

Ellemers, N., R. Spears, and B. Doosje. 2002. Self and social identity. *Annual Review of Psychology* 53, no. 1: 161–86.

Ellemers, N., H. Wilke, and A. Van Knippenberg. 1993. Effects of the legitimacy of low group or individual status on individual and collective status-enhancement strategies. *Journal of Personality and Social Psychology* 64, no. 5: 766–78.

Feick, L., and R.A. Higie. 1992. The effects of preference heterogeneity and source characteristics on ad processing and judgements about endorsers. *Journal of Advertising* 21, no. 2: 9–24.

Feldman, D.C. 1984. The development and enforcement of group norms. *Academy of Management Review* 9, no. 1: 47–53.

Gonzales, A.L., and J.T. Hancock. 2011. Mirror, mirror on my Facebook wall: Effects of exposure to Facebook on self-esteem. *Cyberpsychology, Behavior, and Social Networking* 14, no. 1–2: 79–83.

Gross, R., and A. Acquisti. 2005. Information revelation and privacy in online social networks. In *Proceedings of the 2005 ACM workshop on Privacy in the Electronic Society*, 71–80. New York, NY: ACM. November 2005.

Hampton, K.N., L.S. Goulet, L. Rainie, and K. Purcell. 2011. *Social networking sites and our lives*. Washington, DC: Pew Research Center's Internet & American Life Project.

Han, D., A. Duhachek, and D.D. Rucker. 2015. Distinct threats, common remedies: How consumers cope with psychological threat. *Journal of Consumer Psychology* 25, no. 4: 531–45.

Hennig−Thurau, T., K.P. Gwinner, G. Walsh, and D.D. Gremler. 2004. Electronic word−of−mouth via consumer−opinion platforms: What motivates consumers to articulate themselves on the Internet? *Journal of Interactive Marketing* 18, no. 1: 38–52.

Jackson, L.A., L.A. Sullivan, R. Harnish, and C.N. Hodge. 1996. Achieving positive social identity: Social mobility, social creativity, and permeability of group boundaries. *Journal of Personality and Social Psychology* 70, no. 2: 241–54.

Jetten, J., T. Postmes, and B.J. McAuliffe. 2002. 'We're all individuals': Group norms of individualism and collectivism, levels of identification and identity threat. *European Journal of Social Psychology* 32, no. 2: 189–207.

Kiecker, P., and D. Cowles. 2002. Interpersonal communication and personal influence on the Internet: A framework for examining online word-of-mouth. *Journal of Euromarketing* 11, no. 2: 71–88.

King, H. 2016. Facebook's reach gets even bigger: 1.65 billion monthly users. *CNN Money*, April 27. http://money.cnn.com/2016/04/27/technology/facebook-earnings/index.html

Lalonde, R.N., and R.A. Silverman. 1994. Behavioral preferences in response to social injustice: The effects of group permeability and social identity salience. *Journal of Personality and Social Psychology* 66, no. 1: 78–85.

Laumann, E.O. 1966. *Prestige and association in an urban community*. Indianapolis, IN: Bobbs-Merrill.

Lazarsfeld, P.F., and R.K. Merton. 1954. Friendship as a social process: A substantive and methodological analysis. *Freedom and Control in Modern Society* 18, no. 1: 18–66.

Lee, K.T., M.J. Noh, and D.M. Koo. 2013. Lonely people are no longer lonely on social networking sites: The mediating role of self-disclosure and social support. *Cyberpsychology, Behavior, and Social Networking* 16, no. 6: 413–18.

Luhtanen, R., and J. Crocker. 1992. A collective self-esteem scale: Self-evaluation of one's social identity. *Personality and Social Psychology Bulletin* 18, no. 3: 302–18.

Major, B., and L.T. O'Brien. 2005. The social psychology of stigma. *Annual Review of Psychology* 56: 393–421.

Manago, A.M., M.B. Graham, P.M. Greenfield, and G. Salimkhan. 2008. Self-presentation and gender on MySpace. *Journal of Applied Developmental Psychology* 29, no. 6: 446–58.

Mousavi, S., S. Roper, and K.A. Keeling. 2017. Interpreting social identity in online brand communities: Considering posters and lurkers. *Psychology & Marketing* 34, no. 4: 376–93.

Mummendey, A., and M. Wenzel. 1999. Social discrimination and tolerance in intergroup relations: Reactions to intergroup difference. *Personality and Social Psychology Review* 3, no. 2: 158–74.

Nimrod, G., 2010. Seniors' online communities: A quantitative content analysis. *Gerontologist* 50, no. 3: 382–92.

Nielsen. 2014. *The U.S. digital consumer report*. New York, NY: The Nielsen company. https://www.nielsen.com/content/dam/corporate/us/en/reports-downloads/2014%20Reports/the-digital-consumer-report-feb-2014.pdf

Preacher, K.J., and A.F. Hayes. 2008. Asymptotic and resampling strategies for assessing and comparing indirect effects in multiple mediator models. *Behavior Research Methods* 40, no. 3: 879–91.

Reichelt J., J. Sievert, and F. Jacob. 2014. How credibility affects eWOM reading: The influences of expertise, trustworthiness, and similarity on utilitarian and social functions. *Journal of Marketing Communications* 20, no. 1–2: 65–81.

Rogers, E.M. 1995. *Diffusion of innovations*. New York: Free Press.

Rogers, E.M., and D.K. Bhowmik. 1970. Homophily-heterophily: Relational concepts for communication research. *Public Opinion Quarterly* 34, no. 4: 523–38.

Ruef, M., H.E. Aldrich, and N.M. Carter. 2003. The structure of founding teams: Homophily, strong ties and isolation among U.S. entrepreneurs. *American Sociological Review* 68: 195–222.

Ryu, G., and J.K. Han. 2009. Word-of-mouth transmission in settings with multiple opinions: The impact of other opinions on WOM likelihood and valence. *Journal of Consumer Psychology* 19, no. 3: 403–15.

Schmalz, D.L., C.M. Colistra, and K.E. Evans. 2015. Social media sites as a means of coping with a threatened social identity. *Leisure Sciences* 37, no. 1: 20–38.

Sherman, D.K., and G.L. Cohen. 2006. The psychology of self-defense: Self-affirmation theory. *Advances in Experimental Social Psychology* 38: 183–242.

Smith, A., L. Segall, and S. Cowley. 2012. Facebook reaches one billion users. *CNN Money*, October 4. http://money.cnn.com/2012/10/04/technology/facebook-billion-users/index.html

Steele, C.M. 1988. The psychology of self-affirmation: Sustaining the integrity of the self. *Advances in Experimental Social Psychology* 21, no. 2: 261–302.

Strano, M. 2008. User descriptions and interpretations of self-presentation through Facebook profile images. *Cyberpsychology: Journal of Psychosocial Research on Cyberspace* 2, no. 2, article 5. https://cyberpsychology.eu/article/view/4212

Sussman, S.W., and W.S. Siegal. 2003. Informational influence in organizations: An integrated approach to knowledge adoption. *Information Systems Research* 14, no. 1: 47–65.

Tajfel, H. 1975. The exit of social mobility and the voice of social change: Notes on the social psychology of intergroup relations. *Information (International Social Science Council)* 14, no. 2: 101–18.

Tajfel, H. 1978. *Differentiation between social groups*. London: Academic Press.

Tajfel, H., and J.C. Turner. 1979. *An integrative theory of intergroup conflict*. Monteray, CA: Brooks/Cole.

Tajfel, H., and J.C. Turner. 1986. *The social identity theory of intergroup behavior*. Chicago, IL: Nelson-Hall.

Teng, S., K.W. Khong, W.W. Goh, and A.Y.L. Chong. 2014. Examining the antecedents of persuasive eWOM messages in social media. *Online Information Review* 38, no. 6: 746–68.

Tepper, K. 1994. The role of labeling processes in elderly consumers' responses to age segmentation cues. *Journal of Consumer Research* 20, no. 4: 503–19.

Tesser, A. 2000. On the confluence of self-esteem maintenance mechanisms. *Personality and Social Psychology Review* 4, no. 4: 290–99.

Toma, C.L., and J.T. Hancock. 2013. Self-affirmation underlies Facebook use. *Personality and Social Psychology Bulletin* 39, no. 3: 321–31.

Verkuyten, M. 2005. Ethnic group identification and group evaluation among minority and majority groups: Testing the multiculturalism hypothesis. *Journal of Personality and Social Psychology* 88, no. 1: 121–38.

Verkuyten, M., and A. Reijerse. 2008. Intergroup structure and identity management among ethnic minority and majority groups: The interactive effects of perceived stability, legitimacy, and permeability. *European Journal of Social Psychology* 38, no. 1: 106–27.

Walther, J.B. 2007. Selective self-presentation in computer-mediated communication: Hyperpersonal dimensions of technology, language, and cognition. *Computers in Human Behavior* 23, 5: 2538–57.

Walther, J.B., C. Slovacek, and L.C. Tidwell. 2001. Is a picture worth a thousand words? Photographic images in long term and short term virtual teams. *Communication Research* 28: 105–34.

Wang, Z., J.B. Walther, S. Pingree, and R.P. Hawkins. 2008. Health information, credibility, homophily, and influence via the Internet: Web sites versus discussion groups. *Health Communication* 23, no. 4: 358–68.

White, K., and J.J. Argo. 2009. Social identity threat and consumer preferences. *Journal of Consumer Psychology* 19, no. 3: 313–25.

White, K., J.J. Argo, and J. Sengupta. 2012. Dissociative versus associative responses to social identity threat: The role of consumer self-construal. *Journal of Consumer Research* 39, no. 4: 704–19.

White, K., and D.W. Dahl. 2007. Are all out-groups created equal? Consumer identity and dissociative influence. *Journal of Consumer Research* 34, no. 4: 525–36.

Wilcox, K., and A.T. Stephen. 2013. Are close friends the enemy? Online social networks, self-esteem, and self-control. *Journal of Consumer research* 40, no. 1: 90–103.

Young, K. 2013. Managing online identity and diverse social networks on Facebook. *Webology* 10, no. 2: 1–18.

Zhao, S. 2006. Cyber-gathering places and online-embedded relationships. Paper presented at the annual meetings of the Eastern Sociological Society, February 25, in Boston, U.S.

Zhao, S., S. Grasmuck, and J. Martin. 2008. Identity construction on Facebook: Digital empowerment in anchored relationships. *Computers in Human Behavior* 24, no. 5: 1816–36.

Appendix. Facebook posting stimulus used in studies 1 and 2

 Pat S added 4 new photos
November 16 at 2 17pm

Heard about Brenda's, so drove all the way downtown to check out their dish. Unfortunately, I think people speak too highly about it. After the long hour wait, I really wanted to like this restaurant, but the service was a turn off and the food was just "ok". I'll say, there are infinite options that are way better in the city.

 Like　　Comment　 Share

When brand-related UGC induces effectiveness on social media: the role of content sponsorship and content type

Mikyoung Kim and Doori Song

ABSTRACT
With the growing power of brand-related user-generated content (UGC) on social media, marketers have begun incorporating UGC into the marketing mix as part of word-of-mouth marketing. Drawing upon the Persuasion Knowledge Model, this study examines how content sponsorship interacts with content type to influence consumer responses toward brand-related UGC (inferences of manipulative intent, brand attitude, and intention to click on a URL). The results of an experiment with an online panel in the United States show that when the content is organic (i.e. unpaid), experience-centric content is more likely to induce favourable consumer responses than promotional content. When the content is sponsored (i.e. paid), however, promotional content yields more effective results than experience-centric content. Furthermore, this study demonstrates that consumer inferences of manipulative intent serve as a mediator for the interaction effects between content sponsorship and content types on consumer responses.

Introduction

Brand-related user-generated content (UGC) has become essential product information in consumer purchase decisions (Chu and Kim 2011; Colliander, Dahlen, and Modig 2015; Elwalda, Lu, and Ali 2016). As brand-related UGC has become more influential in the consumer purchase cycle, marketers have used this consumer-to-consumer conversation as a type of marketing practice (Lu, Chang, and Chang 2014). An increasing number of companies are encouraging consumers to spread brand-related information by compensating consumers for so-called sponsored posts (Kozinets et al. 2010). Because of their broad reach, social media have been widely used to implement sponsored posts. One survey result reveals that US social media users read about 86 sponsored posts per month across social media platforms (IZEA 2015). Sponsored brand recommendations in the blogosphere, sponsored stories on Facebook, and sponsored tweets on the microblogging platform Twitter are examples of word-of-mouth (WOM) marketing practices on social media. Among various social media, Twitter, in particular, is a popular venue for spreading sponsored posts due to its far-reaching power in real time (Boerman and Kruikemeier 2016; Chu and Sung 2015). Despite the extensive practice of sponsored posts on Twitter, relatively little is known about consumer responses toward brand-related UGC on Twitter compared to those on other social media such as blogs (Ballantine and Yeung 2015; Hwang and Jeong 2016; Lu, Chang, and Chang 2014). Accordingly, this study attempts to investigate whether and how sponsored brand-related UGC (versus organic) influences consumer responses on Twitter.

Although sponsored brand-related UGC has advantages (e.g. little disruption to users' social media experience), the presence of financial compensation involves risks such as activating consumer persuasion knowledge or increasing the salience of manipulative intent. Consumers' inferences of manipulative intent can undermine the effectiveness of WOM marketing (Boerman and Kruikemeier 2016). Although paid contents are perceived negatively in general, some research notes that sponsored posts on social media can be as effective as other types of marketing tactics (e.g. search engine marketing) (eMarketer 2015). Thus, it is imperative to identify the conditions under which WOM marketing effectiveness is increased. This study proposes the content type of brand-related UGC as a factor contributing to WOM effectiveness. Previous studies on online product reviews have shown that the content type of reviews (factual/objective or evaluative/subjective) influences consumer responses (Jeong and Koo 2015; Kim and Lee 2015). However, little attention has been paid to the effect of content types on consumer responses in WOM marketing research. To fill this gap, this study examines the role of the content type of brand-related UGC (experience-centric and promotional) in consumer responses on Twitter.

The purpose of this study is threefold. Drawing upon the Persuasion Knowledge Model (PKM), this study investigates (a) the main effect of content sponsorship (organic versus sponsored) and (2) the interaction effects between content sponsorship and content type (experience-centric versus promotional) on consumer responses to brand-related UGC (inferences of manipulative intent, brand attitude, and intention to click on a URL in the UGC) on Twitter. Furthermore, this study examines the role of inferences of manipulative intent in the processing of brand-related UGC. The findings of this study contribute to the theoretical advancement of WOM marketing by elucidating the conditions under which the effectiveness of WOM marketing increases on Twitter. This study also provides practical implications for how brand marketers can develop sponsored brand messages and guide WOM agents on social media such as Twitter.

Theoretical background

Brand-related UGC and WOM marketing

WOM has been shared in both offline and online contexts. Due to its independence from marketers' persuasive intentions, this brand-related UGC has been considered more trustworthy and useful in purchase decisions than marketer-generated brand information (Wei and Lu 2013). The power of brand-related UGC has grown as social media penetrates consumers' lives (Kim and Johnson 2016). Since social media enable consumers to create and spread product information with a mere click, even ordinary consumers have a potentially strong influence on brand reputation.

With the exponential power of brand-related UGC, marketers have tried to use these peer-to-peer brand-related conversations for marketing purposes. Marketers reward consumers for spreading positive information about their brands to their social circles (Lu, Chang, and Chang 2014; Wood and Burkhalter 2014). This practice is called WOM marketing, 'the intentional influencing of consumer-to-consumer communications by professional marketing techniques' (Kozinets et al. 2010, 71). Among the various platforms for WOM marketing or sponsored posts, marketers are actively using social media, particularly Twitter for its broad potential reach (Chu, Chen, and Sung 2016). On Twitter, consumers easily pass along brand-related UGC by clicking the 'retweet' button, which spreads brand communications in real time. An example of a WOM marketing practice on Twitter is a sponsored tweet. WOM marketing agencies (e.g. PaidPerTweet) encourage Twitter users to post rewarded brand-related tweets on their own Twitter pages on behalf of the advertiser (Anghelcev 2015). In practice, Twitter users with a large number of followers are hired as WOM agents, but users with a moderate number of followers can work as WOM agents as well. For example, a stay-at-home mom posts a sponsored tweet for Coca-Cola on her Twitter page (e.g. Know someone you want to share a coke with under the mistletoe? Mention their username and tweet with #ShareaCoke! #ad). Twitter users are paid by marketers based either on the number of clicks for their sponsored tweets or on a flat rate per tweet.

Advantages of sponsored UGC relate to its ability to minimize the annoyance of and skeptical attitudes toward brand-related content on social media. Since sponsored (paid) tweets have a form similar to organic (unpaid) tweets, sponsored tweets are less disruptive to consumers' Twitter use than other paid social media advertising (e.g. display ads on Facebook and Twitter) (Campbell and Mark 2015). This organic-like format, however, raises the issue of deception. Due to the format, consumers may be unaware that what they are reading on Twitter is *paid-for* content. For this reason, the Federal Trade Commission (FTC) provides guidelines for sponsored social media content such that any sponsored information should disclose its connection with sponsor organizations (FTC 2013). In the case of sponsored tweets, the identifier 'Ad:' should be included at the beginning of a sponsored tweet to inform consumers that the message is rewarded by a particular advertiser (FTC 2013).

In addition to the issue of deception, the effectiveness of sponsored messages suffers when consumers notice that a brand-related tweet is rewarded by a marketer (Hwang and Jeong 2016). Given this reality, WOM marketers need to understand conditions that alleviate skeptical attitudes toward sponsored brand-related UGC on social media. To find out these conditions, it is important to understand why consumers have scepticism toward sponsored UGC. The process by which consumers develop their responses toward brand-related UGC can be explained by the PKM.

Persuasion Knowledge Model

When consumers encounter brand-related UGC on Twitter, they try to evaluate the usefulness of the information. Consumers' persuasion knowledge enables them to identify an information provider's motive and guides their evaluation of the information. The PKM posits that consumers have knowledge and beliefs about marketers' motives, strategies, and tactics, and they use this knowledge to interpret, evaluate, and respond to persuasion attempts. Persuasion knowledge is developed from personal experiences with persuasion attempts and continues developing throughout life (Friestad and Wright 1994). In general, consumers evaluate familiar and well-developed marketing techniques with confidence due to their accumulated experiences whereas they have difficulty judging new marketing practices with insufficient experiences.

According to the PKM, when consumers perceive a particular message as a persuasion attempt, their persuasion knowledge is activated (Friestad and Wright 1994). Once persuasion knowledge is activated, it leads to suspicion about manipulative intent, 'consumer inferences that the communicator is attempting to persuade by inappropriate, unfair, or manipulative means' (Campbell 1995, 228). This inference of manipulative intent induces resistance to the persuasive attempt, resulting in less favourable brand evaluations (Boerman and Kruikemeier 2016; Campbell 1995; Campbell and Kirmani 2000; Wei, Fischer, and Main 2008). For example, Boerman and Kruikemeier (2016) found that consumers reacted less favourably toward a promoted tweet from a political party than toward a non-promoted tweet due to the activated persuasion knowledge.

Applying the PKM, we propose that consumers are likely to use their persuasion knowledge and to try to infer motives behind the messages. One condition that influences the activation of persuasion knowledge is sponsorship of brand-related UGC, which varies in terms of the salience of manipulative intent.

Content sponsorship

Peer-to-peer brand-related communication occurs naturally. Consumers share their brand experiences in person or on the Internet. With the growing power of consumer brand-related conversations, however, some consumers post their brand-related opinions for material rewards (e.g. money, free product samples, and gift cards). On social media, brand-related UGC can be classified as either organic or sponsored. Organic brand-related UGC is unpaid and includes consumers' genuine brand opinions. By contrast, sponsored brand-related UGC is guided and paid for by marketers (Hwang and Jeong 2016). On Twitter, companies pay for sponsored tweets (Wood and Burkhalter 2014).

As paid content permeates the realm of social media, a tension occurs because the way WOM marketing works (i.e. compensating consumers in return for spreading brand information) collides with what consumers expect on social media. People are likely to perceive social media as their space rather than marketers'. By nature then, sponsored UGC may draw more negative consumer responses than its organic counterpart. Consistent with this reasoning, previous studies in the context of WOM marketing have shown the positive effect of organic content (versus sponsored) on brand attitudes (Wei et al. 2008) and behavioural intentions (Wood and Burkhalter 2014). For example, Wood and Burkhalter (2014) found that consumers who received an organic brand-related tweet were more likely to click on the URL in the tweet than were those who received a sponsored brand-related tweet. The superior effects of organic brand-related UGC (versus sponsored) on consumer responses can be attributed to consumers' tendencies to infer information providers' motives (Friestad and Wright 1994; Lee, Kim, and Peng 2013; Reeder 2009). Inferences of motives are guided by how consumers perceive persuasion attempts, as obvious and appropriate or as manipulative (Campbell 1995; Wei et al. 2008), and these inferences, in turn, influence the effectiveness of information (e.g. brand attitude, purchase intention, etc.).

In the context of brand-related UGC on Twitter, a motive inference for organic UGC seems to be relatively simple. Since an information provider has no economic gain for sharing brand-related information, organic brand-related UGC will be perceived as unbiased and appropriate. Thus, consumers may deduce genuine motives and have favourable responses toward organic content. By contrast, the presence of financial rewards in sponsored brand-related UGC is perceived as biased and inappropriate in the context of UGC, and hence activates consumer persuasion knowledge. This activated persuasion knowledge guides consumers' inferences of the motives behind sponsored UGC. For sponsored UGC, multiple plausible causes exist: brand-related UGC is posted based on (1) an information provider's intention to help others; (2) compensation by marketers; (3) or both. According to Kelley's discounting principle (1973), 'the role of a given cause in producing a given effect is discounted if other plausible causes are also present' (113). If the observed effect (e.g. brand-related UGC) is explained by an external justification (e.g. compensation), an internal cause (e.g. the communicator's altruism) is discounted (Kelley 1973). Applying the PKM and the discounting principle, this study expects that consumers will infer manipulative intent in sponsored brand-related UGC, resulting in less-favourable responses. Thus, we hypothesize:

H1a. Organic UGC will lead to less inference of manipulative intent than sponsored UGC.
H1b. Organic UGC will lead to more favourable brand attitude than sponsored UGC.
H1c. Organic UGC will lead to greater intention to click on a URL in the UGC than sponsored UGC.

Content sponsorship and content type

Although organic UGC is expected to prove superior to sponsored UGC on Twitter, the effect of content sponsorship on consumer responses may not be uniform. Another content-related factor that influences consumer responses to brand-related UGC is content type (Jeong and Koo 2015; Kim and Lee 2015). This study expects an interaction effect between content sponsorship and content types on consumer responses to brand-related UGC.

Content type refers to the features of brand-related UGC provided by content creators in this study. On Twitter, several types of brand-related content are shared: descriptive, experience-centric, comparative, and promotional (Lee et al. 2014). Descriptive messages include a simple description of the brand. Experience-centric messages contain consumers' personal experiences and opinions about the brand. Comparative messages include comparisons between brands. Finally, promotional messages include a webpage or a link to introduce other content. Among these various types of brand-related content, the two types that this study focuses on are experience-centric and promotional due to their prevalence on Twitter. First, experience-centric content involves consumers' subjective opinions about a particular product or service based on their personal experiences (Huang et al. 2015). Experience-centric information consists of an overall product evaluation (good or bad) and a written explanation for the evaluation (e.g. product attributes) (Lee, Park, and Han 2008). An example of experience-centric content on Twitter is: 'I love using #MountainHighYogurt as a substitute in my baking! It's tasty.' Traditional WOM and consumer review messages contain experience-centric brand information. Experience-centric information accounts for a major share of brand-related UGC on social media (Lee et al. 2014).

Another UGC type on Twitter is promotional (e.g. 'Check out Craftsy's Memorial Day Sale! Enjoy up to 70% off all kits, fabric, and yarn, up to 60% off all'). Consumers share promotional information on social media to satisfy their desires for self-fulfilment (Fu, Wu, and Cho 2017). Unlike experience-centric content, promotional content consists of factual and objective information such as potential economic benefits (e.g. price discounts and coupons) (Vermeir and Van Kenhove 2005). Although promotional information has not been common in the context of WOM, promotions have been shared recently on social media. As an increasing number of companies use social media as a major venue for digital advertising, promotional information has been posted on brands' social media pages (Schultz and Peltier 2013). Due to its perceived benefits, consumers search actively and share promotional information on the Internet. For example, searching information on social discounting platforms (e.g. Groupon) and social commerce sites (e.g. GasBuddy) is routine among tech-savvy consumers (Edelman, Jaffe, and Kominers 2016), and gaining economic benefits (e.g. coupons and promotions) is one motive for consumers to join brand communities on social media (Sung et al. 2010).

On Twitter, consumers may perceive both experience-centric and promotional content as beneficial information. Experience-centric content reduces consumers' information search costs and perceived risk regarding products (Goldsmith and Horowitz 2006). Before purchasing products, consumers can gain indirect knowledge of products via other consumers' experience-centric content. On the other hand, promotional content offers economic benefits by lowering the cost of products (Chandon, Wansink, and Laurent 2000). Although both content types offer benefits to consumers, an important question is whether these perceived benefits will change the effect of content sponsorship on consumer responses to brand-related UGC.

From the PKM standpoint, sponsored brand-related UGC is more likely to be associated with persuasive intent. The sponsorship label (#ad or #spon) works as a trigger that activates consumer persuasion knowledge (Boerman and Kruikemeier 2016) and this activated persuasion knowledge induces unfavourable responses. However, we expect that the detrimental effect of sponsored UGC can be mitigated by adopting the appropriate type of content. According to the economics of information, consumers are more skeptical of difficult-to-verify persuasive messages and less skeptical of easily verifiable messages (Ford, Smith, and Swasy 1990). For example, consumers are more skeptical toward advertising claims when they are not able to determine the product quality prior to purchase (i.e. experience qualities) than when they can (i.e. search qualities). Between experience-centric and promotional content, the content of promotional information is easily verified because it includes factual and objective information. By contrast, consumers may have difficulty verifying experience-centric content because it contains personal and subjective opinions which are susceptible to erroneous interpretations. Consequently, consumers may have more skeptical attitudes toward experience-centric content than toward promotional content. Taking the economics of information perspective together with the PKM, we predict that when consumers notice financial rewards, they might not invest their cognitive efforts to infer an information provider's motive for promotional content because it includes only factual and objective information about a product or service. On the other hand, with sponsored experience-centric content, consumers face a complicated situation in inferring an information provider's motive as either altruistic or monetary, or both. These multiple plausible causes make it difficult for consumers to verify the message content and will induce inferences of manipulative intent and unfavourable responses.

Meanwhile, the positive effect of organic UGC on consumer responses will be pronounced with experience-centric content. Since experience-centric content is a common form of traditional WOM and consumer reviews, organic experience-centric UGC will not activate consumer persuasion knowledge on Twitter. However, promotional content (versus experience-centric content) may be perceived as inappropriate on social media because it looks like a marketing message. This marketing-like information can activate consumer persuasion knowledge and induce inferences of manipulative intent and less-favourable consumer responses.

Based on the above discussion, this study proposes that for organic UGC, experience-centric content will be less likely to draw manipulative intent and will generate more favourable responses than promotional content. For the sponsored UGC, however, the promotional content will be less likely to induce manipulative intent and will lead to more favourable responses than experience-centric content. Thus, we hypothesize:

H2a. When brand-related UGC is organic, experience-centric content will lead to less inferences of manipulative intent than promotional content, whereas when brand-related UGC is sponsored, promotional content will lead to less inferences of manipulative intent than experience-centric content.

H2b. When brand-related UGC is organic, experience-centric content will lead to more favourable brand attitude than promotional content, whereas when brand-related UGC is sponsored, promotional content will lead to more favourable brand attitude than experience-centric content.

H2c. When brand-related UGC is organic, experience-centric content will lead to greater intention to click on a URL than promotional content, whereas when brand-related UGC is sponsored, promotional content will lead to greater intention to click on a URL than experience-centric content.

Mediating role of inferences of manipulative intent

Previous research suggests that inferences of manipulative intent influence subsequent brand evaluations (Campbell 1995; Campbell and Kirmani 2000; Hibbert et al. 2007; Lunardo and Mbengue 2013). For example, Campbell (1995) found that inferences of manipulative intent mediated the relationship between various attention-getting tactics (i.e. personal investment, personal benefits, and advertiser investment) and brand evaluations (i.e. attitude toward the ad, attitude toward the brand, and purchase intentions). Similarly, in a personal sales context, Campbell and Kirmani (2000) found that the salience of a salesperson's persuasive intent and a consumers' cognitive capacity influenced consumer inferences of ulterior (manipulative) motives, and these inferences, in turn, affected perceptions of a salesperson's sincerity. This study examines whether inferences of manipulative intent is an underlying mechanism that can explain the interaction effect of content sponsorship and content type on consumer responses toward brand-related UGC. Thus,

H3a. Inferences of manipulative intent will mediate the interaction effect of content sponsorship and content type on brand attitude.

H3b. Inferences of manipulative intent will mediate the interaction effect of content sponsorship and content type on intention to click on an URL in the UGC.

Method

To test the proposed hypotheses, a 2 (content sponsorship: organic versus sponsored) × 2
(content type: experience-centric versus promotional) between-subjects experimental
design was employed in the Twitter context. All variables were manipulated.

Stimuli development

To select the appropriate product category, a pretest with 49 participants (different from
the main study's participants) was conducted. Participants were given a list of six different
product categories[1] (pet supplies, herbal remedy products, fitness centres, hard drives,
eye glasses, and electric toothbrushes) and asked to indicate their level of product
involvement using 10 seven-point scale items (Zaichkowsky 1994). Based on the pretest
results, an electric toothbrush was chosen for this study ($M_{pet\ supplies}$ = 4.72; $M_{herbal\ remedies}$
= 4.12; $M_{fitness\ centre}$ = 5.12; $M_{hard\ drive}$ = 4.65; $M_{glasses}$ = 4.73; $M_{electric\ toothbrush}$ = 5.26). A fictitious brand name (Lumen) was used to control for pre-existing attitudes toward the brand
(Till and Busler 2000).

To manipulate content sponsorship, two versions of brand-related UGC were created:
organic and sponsored. Both versions were identical except for the presence of the disclosure of content sponsorship. Specifically, the sponsored brand-related UGC contained a
disclosure term (#ad) at the end of the content, whereas the organic brand-related UGC
did not contain any disclosure term.

Since WOM marketing is a fairly new marketing practice, many consumers may not recognize sponsored UGC on Twitter. Indeed, only 20% of the participants recognized sponsored tweets in the study of Boerman and Kruikemeier (2016). However, it was critical that
participants were knowledgeable about sponsored brand-related UGC on Twitter and
could discern sponsored UGC from organic UGC to examine the effect of content sponsorship on consumer responses toward the UGC in this study. To help consumers identify the
content sponsorship of tweets, all participants were asked to read a fictitious trade journal
article regarding sponsored tweets before reading the brand-related UGC. The article
included information about what a sponsored tweet is, how sponsored tweets work, and
how consumers can distinguish sponsored from organic tweets.

To manipulate the content type, two versions of brand-related UGC were developed.
Experience-centric content consists of an overall product evaluation (good or bad) and an
explanation for that evaluation, such as opinions about product attributes and performance
(Burns, Temkin, and Geller 2008). To choose the attributes of an electric toothbrush to be
included in the brand-related UGC, various websites containing consumer product reviews
about electric toothbrushes (e.g. Amazon.com) were reviewed. The three most frequently
mentioned attributes in consumer product reviews were gentleness, ease of use, and battery
life. These three attributes were included in the target brand-related UGC as shown below.

> My new Lumen electric toothbrush is great! It's gentle on my teeth and gums, easy to use, and
> has long battery life. Check it out! http://bit.ly.lt

On the other hand, the promotional content contained information about economic benefits such as free shipping and a discounted price as shown below.

I found a great deal for Lumen electric toothbrushes! Lumen's got free shipping & 25% off with promo code L25. Check it out! http://bit.ly.lt

Four versions of a Twitter page were created by modifying existing Twitter pages. Except for the manipulations of content sponsorship (organic or sponsored) and content type (experience-centric or promotional), the four pages were identical in terms of page layout and seven non-brand related tweets (e.g. 'I love my life! Wish you all the same!').

Participants and procedures

Overall, 130 adult Twitter users in the United States participated in this study. Participants were recruited from Survey Sampling International (SSI). Among the panel, this study's sample was limited to (1) Twitter users and (2) those aged 18–34.[2] Each participant received 300 points (the equivalent of $3) in return for this study. The average sample age was 26.8 (SD = 3.97) with ages ranging from 19 to 34. Females (56%) outnumbered males (44%). Most participants were white (63.1%), while the rest were black (14.6%), Hispanic (8.5%), Asian (8.5%), and other (5.3%). More than half of the participants (55%) had earned at least a university or advanced degree. Participants, on average, had used Twitter for more than one year (M = 17.54 months, SD = 9.79 months). More than half of the participants (56%) reported spending less than an hour on Twitter each day (for more details, see Table 1).

Table 1. Sample characteristics (n = 121).

		Percentage or Mean
Gender	Male	44
	Female	56
Age	19–24	28.5
	25–30	50.8
	Over 30	20.7
Ethnicity	Asian	8.5
	Black or African American	14.6
	Hispanic	8.5
	White	63.1
	Others	5.3
Education level	Senior high school or below	45.3
	Bachelor/college degree	50
	Master degree or above	4.7
Income level	Less than $20,000	13.2
	$20,000–$39,999	19.4
	$40,000–$59,999	25.6
	$60,000–$79,999	17.8
	$80,000–$99,999	9.3
	$100,000 or more	14.7
Twitter usage	Usage period	17.54 months
	Usage Hours	
	Less than 1 hour per	56
	1 hour to less than 2 hours	19.2
	2 hours to less than 3 hours	14.5
	3 hours to less than 4 hours	3.6
	4 hours to less than 5 hours	2.6
	More than 5 hours	4.1

The experiment was conducted online. Participants were given the experiment website URL. All participants were first asked to read a fictitious trade journal article about sponsored tweets. Immediately after reading the article, participants were asked to answer questions testing their knowledge about sponsored tweets.

On the next page, Java Script directed participants randomly to one of four conditions. After reading the Twitter page, they were asked to complete the remaining parts of the questionnaire including dependent measures, manipulation checks, and demographic information.

Measures

Inferences of manipulative intent were measured by asking participants to indicate their level of agreement or disagreement with each of the following five statements on a seven-point scale: 'The way the information provider tried to persuade me seems acceptable to me,' 'The information provider tried to manipulate people in ways that I do not like,' 'I was annoyed by the information provider's tweets about products because s/he seemed to be trying to inappropriately manage or control people,' 'I do not mind the information provider's tweets about products; the information provider tried to be persuasive without being excessively manipulative,' and 'The information provider's tweets about products were fair in what was said and shown' (Campbell 1995) ($\alpha = .90$).

Attitude toward the brand was measured with four seven-point semantic differential items anchored by 'dislike/like,' 'negative/positive,' 'unfavourable/favourable,' and 'bad/good' (Holbrook and Batra 1987) ($\alpha = .96$).

Intention to click on a URL in the UGC was measured with four seven-point items anchored by 'unlikely/likely,' 'improbable/probable,' 'impossible/possible,' and 'uncertain/certain' (Bearden, Lichtenstein, and Teel 1984) ($\alpha = .97$).

To control the impact of product involvement, participants were asked to indicate their level of product involvement with 10 seven-point semantic differential items (e.g. unimportant/important) (Zaichkowsky 1994) ($\alpha = .92$).

For subsequent analyses, scores of all items for each construct were averaged.

Results

Manipulation check

For the manipulation check for content sponsorship, participants were asked to answer whether the targeted Twitter message on the electric toothbrush was sponsored or not, with three options (True/False/Don't know). Out of 130 participants, 81% perceived the organic UGC condition correctly and 84% perceived the sponsored UGC condition correctly, $\chi^2 = 56.68$, $p < .001$. To increase the accuracy of the experiment, participants who answered the assigned content sponsorship (organic or sponsored) incorrectly were excluded from further analyses. As a result, 121 responses were included in the final analyses. To confirm the success of the content type manipulation, participants were asked to rate the experiential or promotional value of the review (1 = entirely experience-centric, 7 = entirely promotional). As intended, an independent t-test showed that participants exposed to experience-centric UGC perceived the information as more experiential

(M = 2.27, SD = 1.02) than did those exposed to promotional UGC (M = 5.47, SD = 1.17), t (119) = −15.92, $p < .001$. These results show that the manipulation of the content sponsorship and content type were successful.

Hypotheses testing

To test the stated hypotheses, a Multivariate Analysis of Covariance (MANCOVA) with product involvement as a covariate was conducted. The MANCOVA revealed a significant multivariate effect of product involvement, Wilk's λ = .43, F (3, 114) = 51.51, $p < .001$, a significant main effect of content sponsorship, Wilk's λ = .59, F (3, 114) = 26.88, $p < .001$, and a significant two-way interaction effect between sponsorship and content type, Wilk's λ = .88, F (3, 114) = 5.12, $p < .01$. Content type did not have a significant main effect on dependent variables, Wilk's λ = .95, F (3, 114) = 1.96, n.s.

This study predicted the main effects of content sponsorship on inferences of manipulative intent (H1a), brand attitude (H2b), and click intention (H1c). Univariate analyses revealed significant main effects on inferences of manipulative intent, F (1, 116) = 31.71, $p < .001$, brand attitude, F (1, 116) = 10.17, $p < .01$, and click intention, F (1, 116) = 65.89, $p < .001$. Specifically, consistent with H1a, a pairwise comparison test indicated that organic UGC drew less inferences of manipulative intent ($M_{organic}$ = 2.78, $M_{sponsored}$ = 3.94, $p < .001$). Further, in support of H1b, a pairwise comparison test indicated that participants exposed to organic UGC (M = 4.81) had more favourable brand attitude than those exposed to sponsored UGC (M = 4.32) ($p < .01$). Finally, as expected, a pairwise comparison test indicated that organic UGC (M = 4.89) led to greater intention to click on a URL than sponsored UGC (M = 3.28) ($p < .001$). Thus, H1a–c were supported.

H2a–c posited the interaction effects of content sponsorship and content type on inferences of manipulative intent (H2a), brand attitude (H2b), and click intention (H2c). A univariate analysis revealed a significant two-way interaction effect on inferences of manipulative intent, F (1, 116) = 13.32, $p < .001$ (see Figure 1). As expected, the contrast tests indicated that when the brand-related UGC was organic, the experience-centric

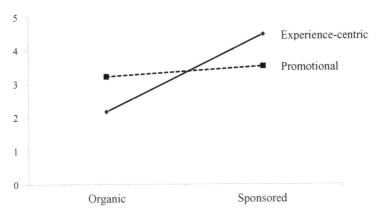

Figure 1. Interaction effect between content sponsorship and content type on inference of manipulative intent.

content (M = 2. 17, SD = 1.33) drew less inferences of manipulative intent than the promotional content (M = 3.22, SD = 1.17), F (1, 116) = 4.98, p < .05. When the brand-related UGC was sponsored, however, the promotional content (M = 3.53, SD = .96) drew less inferences of manipulative intent than the experience-centric content (M = 4.48, SD = 1.57), F (1, 116) = 8.99, p < .01. Thus, H2a was supported.

A significant two-way interaction effect on brand attitude was revealed, F (1, 116) = 6.17, p < .05 (see Figure 2). The contrast tests indicated that for organic UGC, the experience-centric content (M = 5.44, SD = 1.03) drew more favourable brand attitude than the promotional content (M = 4.37, SD = 1.01), F (1, 116) = 13.48, p < .001. However, for sponsored UGC, the levels of brand attitude were similar between the experience-centric (M = 4.14, SD = 1.41) and the promotional content (M = 4.44, SD = 1.00), F (1, 116) = 1.10, n.s. Thus, H2b was partially supported.

In support of H2c, a univariate analysis revealed a significant two-way interaction effect on click intention, F (1, 116) = 4.89, p < .05, (see Figure 3). The contrast tests indicated that when the brand-related UGC was organic, the experience-centric content (M = 5.49, SD = 1.24) led to greater intention to click on a URL than the promotional content

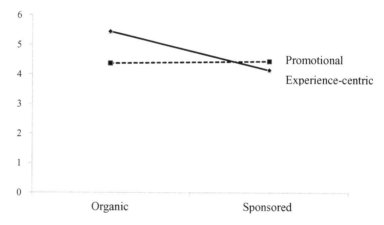

Figure 2. Interaction effect between content sponsorship and content type on brand attitude.

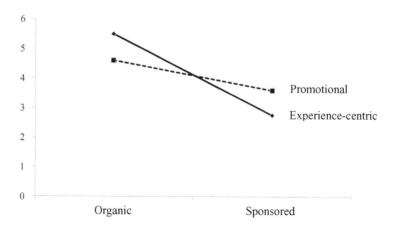

Figure 3. Interaction effect between content sponsorship and content type on click intention.

Table 2. Descriptive statistics.

	Organic		Sponsored	
	Experience (n = 26)	Promotional (n = 31)	Experience (n = 32)	Promotional (n = 32)
Inferences of manipulative intent	2.17 (1.33)	3.22 (1.17)	4.48 (1.57)	3.53 (.96)
Brand attitude	5.44 (1.03)	4.37 (1.01)	4.14 (1.41)	4.44 (1.00)
Intention to click on a URL	5.49 (1.03)	4.60 (1.39)	2.75 (1.65)	3.59 (1.76)

Note: Standard deviations are in parentheses.

(M = 4.60, SD = 1.39), F (1, 116) = 4.80, p < .05. When the brand-related UGC was spon-sored, however, the promotional content (M = 3.59, SD = 1.76) yielded greater intention to click on a URL than the experience-centric content (M = 2.75, SD = 1.65), F (1, 116) = 4.84, p < .05. Thus, H2c was supported. Table 2 includes descriptive statistics.

H3a–b predicted the mediating role of manipulative intent in the two-way interaction effect on brand attitude (H3a) and click intention (H3b). To test this moderated mediation hypothesis, a conditional indirect effect analysis (Preacher, Rucker, and Hayes 2007) was performed using the MODMED SPSS macro. The results indicated a significant interaction effect of the content sponsorship and the content type on inferences of manipulative intent (b = −2.00, SE = .47, t = −4.29, p < .001). Consistent with H3a and 3b, the inferences of manipulative intent, in turn, influenced brand attitude (b = −.48, SE = .07, t = −7.23, p < .001) and click intention (b = −.62, SE = .09, t = −6.42, p < .001) (see Figure 4). Specifically, the indirect effect of the experience-centric content (versus promotional) on brand atti-tude (indirect effect = .54, SE = .19, p < .01) and click intention (indirect effect = .65, SE = .23, p < .01) via inferences of manipulative intent was significant for organic UGC. How-ever, for sponsored UGC, the promotional content (versus experience-centric) had a signif-icant effect on brand attitude (indirect effect = −.49, SE = .18, p < .01) and intention to click (indirect effect = −.59, SE = .22, p < .01) via inferences of manipulative intent. Thus, H3a and H3b were supported (see Table 3 for moderated mediation results).

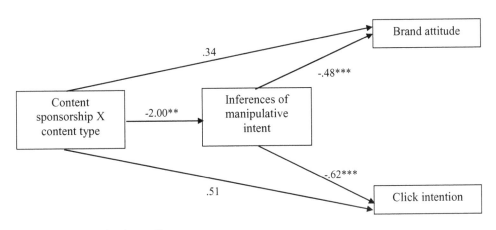

Figure 4. Conditional indirect effects.
***p < .001, **p < .01, *p < .05.

Table 3. Moderated mediation results.

	Coefficient	SE	T	p
DV = Inferences of manipulative intent				
Content sponsorship	0.95	0.32	2.98	0.004
Content type	−0.31	0.32	−0.95	0.34
CS × CT	−2	0.47	−4.29	0
DV = Brand attitude				
Inferences of manipulative intent	−0.48	0.07	−7.23	0
CS × CT	0.34	0.37	0.94	0.35
DV = Click intention				
Inferences of manipulative intent	−0.62	0.09	−6.42	0
CS × CT	0.51	0.52	0.97	0.33
Conditional indirect effect				
Dependent variable	Content sponsorship	Indirect effect	SE	p
Brand attitude	Organic	0.54	0.19	0.004
	Sponsored	−0.49	0.18	0.006
Click intention	Organic	0.65	0.23	0.006
	Sponsored	−0.59	0.22	0.007

Note. Bold values represent those used to test hypotheses.

Discussion

As social media penetrates consumers' lives, marketers try to find ways to use social media smartly. One such attempt is WOM marketing practices (Berger 2014). However, since WOM marketing is a relatively new marketing practice, knowledge about efficient WOM marketing implementation is insufficient. Therefore, the need exists to identify conditions which minimize consumer scepticism and maximize persuasion effects. In response, employing the PKM, the purpose of this study was threefold. First, this study investigates whether and how consumer responses to organic and sponsored brand-related UGC (inferences of manipulative intent, brand attitude, and intention to click on a URL in the UGC) on Twitter differ. Second, this study examines a boundary condition (i.e. content type) that influences the effect of content sponsorship on consumer responses. Finally, this study investigates whether inferences of manipulative intent work as an underlying mechanism to explain the effects of brand-related UGC on consumer responses.

Summary of findings

The results reveal that organic (unpaid and natural) brand-related UGC induces fewer inferences of manipulative intent and more favourable responses than sponsored (paid) UGC on Twitter. This finding coincides with the results of previous studies in WOM marketing such that an unpaid message is more effective than a paid message (Anghelcev 2015; Boerman and Kruikemeier 2016; Tuk et al. 2009; Wood and Burkhalter 2014; Wei et al. 2008).

The key findings of this study are interaction effects of content sponsorship and content type on consumer responses and a mediating role of inferences of manipulative intent. For organic (unpaid) brand-related UGC, experience-centric content yields more favourable brand attitude and greater intention to click on a URL in the UGC than promotional content on Twitter. Interestingly, for sponsored brand-related UGC, promotional content yields greater intention to click on a URL in the UGC than experience-centric content.

These findings support the predictions of this study. In accordance with the PKM, the content of organic UGC is perceived as unbiased (Boerman and Kruikemeier 2016; Wood and Burkhalter 2014). Given this condition, experience-centric content, a common form of WOM, is likely to be blended with organic UGC. The familiar form of organic experience-centric UGC is perceived as appropriate and natural on Twitter and thus may not activate consumer persuasion knowledge. However, since promotional information, taking after traditional marketing messages, has been an uncommon type in the UGC context, it may be perceived as inappropriate and thus activate consumer persuasion knowledge even if it is organic. This activated persuasion knowledge leads to inferences of manipulative intent and unfavourable consumer responses (Reeder 2009).

On the other hand, sponsored UGC is likely to evoke persuasion knowledge. Given the activation of persuasion knowledge, consumers may look for another cue to infer a motive. Applying the economics of information (Ford, Smith, and Swasy 1990), the quality of promotional content is easily verified, whereas the verification of experience-centric content is difficult. Together with the presence of financial rewards, consumers may infer multiple motives for unverifiable experience-centric content. The presence of an external cause (financial rewards) may discount internal causes (altruistic motives) as Kelley's discounting principle (1973) suggested, boost inferences of manipulative intent, and lower intention to click on a URL in the sponsored experience-centric UGC.

Unlike our prediction, however, for sponsored brand-related UGC, the level of brand attitude was similar regardless of content type on Twitter. This finding indicates that content type (experience-centric or promotional content) can be an effective boundary condition in increasing the intention to click on a URL for the sponsored brand-related UGC, but it has minimal impacts on forming brand attitudes. Future research needs to examine whether there is any condition that sponsored contents influence consumers' brand attitudes on social media.

Further, this study's results reveal that consumer inferences of manipulative intent play a significant mediating role between two-way interaction effects and consumer responses. The significant mediating role of inferences of manipulative intent in this study converges with findings of previous studies on advertising tactics (Campbell 1995; Hibbert et al. 2007), sales contexts (Campbell and Kirmani 2000), retail contexts (Lunardo and Mbengue 2013), and corporate social responsibility (CSR) (Forehand and Grier 2003).

Theoretical implications

This study's results suggest several theoretical implications. Theoretically, this research advances WOM marketing literature by (1) identifying factors influencing WOM marketing effectiveness on Twitter and (2) identifying a cognitive mechanism underlying WOM marketing effectiveness. First, this study's findings add insight into WOM marketing by identifying influential conditions of WOM marketing. Most previous studies on WOM marketing have explored the role of the disclosure of rewards (Hwang and Jeong 2016; Lu, Chang, and Chang 2014), reward types (Anghelcev 2015; Lu et al. 2014), brand familiarity (Wei et al. 2008; Wood and Burkhalter 2014), and sources of information (Boerman and Kruikenmeier 2016; Wood and Burkhalter 2014). However, little attention has been paid to the content aspect of information, although content is a strong driver for perceived usefulness (Kim and Lee 2015). By incorporating content type (experience-centric versus promotional) and content sponsorship in brand-related UGC, this study contributes to the identification of the conditions that mitigate the detrimental effects of sponsored UGC on Twitter.

Second, this study advances WOM marketing literature by identifying a role of inferences of manipulative intent. Previous studies on WOM marketing have paid little attention to consumer inferences of manipulative intent as a possible underlying mechanism in the effectiveness of brand-related UGC. However, virtually every marketing message includes claims that are subject to some degree of disbelief (Obermiller and Spangenberg 2000) and that cause consumers to try to cope with marketers' persuasion attempts (Friestad and Wright 1994). Employing the PKM, this study reveals that the effect of unpaid content is maximized with experience-centric content, whereas the effect of paid and strategic content is fostered with promotional content because the information providers' motives are not perceived as manipulative.

Practical implications

The findings of this study provide practical insights for digital marketing practitioners into how to use social media as marketing communication. First, it is better for marketers to facilitate natural peer-to-peer brand conversations rather than manage brand information flow by offering financial rewards. Consistent with this suggestion, the warranting theory notes that, because of perceived objectivity and credibility, information from peers has greater influence on consumers' perceptions and behaviours than marketer-generated information (Walther et al. 2009). Marketers who engage in WOM marketing on Twitter try to provide interesting and valuable content on their social media pages to encourage voluntary retweeting and pass-along behaviour.

Second, our results suggest that when marketers implement sponsored posts, they should have promotional content. Since organic WOM marketing and online reviews generally consist of experience-centric content, consumers perceive such content as inappropriate and deceitful when they recognize that the content is sponsored. In contrast, consumers are less likely to perceive promotional content as deceitful even when they are aware of the presence of rewards. As the economics of information suggest, consumers give more credence to information when the information cannot be manipulated easily by message providers (Obermiller and Spangenberg 2000). Consumers are less likely to make inferences of manipulative intent toward promotional content (objective information). Further, intention to click on a URL for sponsored promotional UGC is greater than that for sponsored experience-centric UGC on Twitter. This is particularly true for unfamiliar brands. Thus, a promotional message can be a good way to spread sponsored brand-related information for new or unfamiliar brands on Twitter.

The findings of this study provide implications for policymakers. Although a substantial number of participants distinguished organic from sponsored content in this study (81% for organic and 84% for sponsored), about 15% of participants exposed to the sponsored content were unaware of content sponsorship despite sponsored tweet-related information given in advance. The recognition rate of content sponsorship is seriously low when information about sponsorship is not offered. Only 20% of the participants noticed the sponsorship label in Boerman and Kruikemeier's study (2016). It is uncertain how many consumers understand the meaning of #ad or #spon on social media in real situations. These results may encourage policymakers to reinforce disclosure guidelines as well as to develop a program for educating consumers about sponsored content on social media.

Limitations and future research

While this study contributes to persuasion literature and social media marketing practices, future studies are called for to overcome its limitations. First, several issues relate to its external validity. The emergence of Internet-centric social media has caused consumers to seek product information from various forms of consumer-generated media including micro blogs (e.g. Twitter), video-sharing platforms (e.g. YouTube), and image-sharing sites (e.g. Instagram). This study examines proposed hypotheses pertinent to only one social media platform: Twitter. Given that companies employ various types of social media, it is worthwhile to examine how different types of social media have different effects on each stage of the persuasion process with considerations of sponsorship as well as types of content. For example, social networking sites (e.g. Facebook) are good places for building consumer-brand relationships (Kim, Sung, and Kang 2014), whereas user-sponsored blogs (e.g. the unofficial Apple weblog and CNET.com) may be outlets for informing and spreading company-related information. Future research needs to investigate how the characteristics of each social medium contribute to WOM marketing effectiveness. Another external validity issue relates to the presence of organic brand-related UGG with promotional content. Although consumers share brand-related promotional messages with their close friends on social media platforms (Choi, Seo, and Yoon 2017), they are less likely to post organic brand-related UGC with promotional content than organic UGC with experience-centric content. When considering the fact that organic promotional UGC is not prevalent, the observed effects may be derived from the oddity of such contents in this study.

Third, this study's tested product category, an electronic toothbrush, limits the generalizability of the findings. Since previous studies found that product type influences brand evaluations (Lu, Chang, and Change 2014), future studies need to examine whether this study's findings can be replicated with experience goods.

Finally, although the providers of brand-related UGC in this study were established as ordinary consumers, previous studies have suggested that the effects of the same message can be different depending on the author or re-poster (Boerman and Kruikemeier 2016; Halliday 2016; Jin and Phua 2014; Paek et al. 2011; Wood and Burkhalter 2014). On social media, influencers such as celebrities or experts have differing information-spreading power and perceived credibility. It would be valuable, interesting, and possibly even revelatory, to examine that aspect also.

Notes

1. The list of product categories was developed based on the analysis of 500 sponsored tweets at the point of data collection.
2. The largest age group among US Twitter users is young adults 18–34 (http://www.quantcast. com).

Disclosure statement

No potential conflict of interest was reported by the authors.

Funding

This work was supported by 2016 Hongik University Research Fund.

References

Anghelcev, G. 2015. Unintended effects of incentivizing consumers to recommend a favorite brand. *Journal of Marketing Communications* 21: 210–23.

Ballantine, P.W., and C.A. Yeung. 2015. The effects of review valence in organic versus sponsored blog sites on perceived credibility, brand attitude, and behavioral intentions. *Marketing Intelligence & Planning* 33: 508–21.

Bearden, W.O., D.R. Lichtenstein, and J.E. Teel. 1984. Comparison price, coupon, and brand effects on consumer reactions to retail newspaper advertisements. *Journal of Retailing* 60: 11–34.

Berger, J. 2014. Word of mouth and interpersonal communication: A review and directions for future research. *Journal of Consumer Psychology* 24: 586–607.

Boerman, S.C., and S. Kruikemeier. 2016. Consumer responses to promoted tweets sent by brands and political parties. *Computers in Human Behavior* 65: 285–94.

Burns, M., B. Temkin, and S. Geller. 2008. The design guide for online customer reviews: A social computing report. https://www.forrester.com/The+Design+Guide+For+Online+Customer+Reviews/fulltext/-/E-res44373.

Campbell, M.C. 1995. When attention-getting advertising tactics elicit consumer inferences of manipulative intent: The importance of balancing benefits and investments. *Journal of Consumer Psychology* 4: 225–54.

Campbell, M.C., and A. Kirmani. 2000. Consumers' use of persuasion knowledge: The effects of accessibility and cognitive capacity on perceptions of an influence agent. *Journal of Consumer Research* 27: 69–83.

Campbell, M.C., and L.J. Marks. 2015. Good native advertising isn't a secret. *Business Horizons* 58: 599–606.

Chandon, P., B. Wansink, and G. Laurent. 2000. A benefit congruency framework of sales promotion effectiveness. *Journal of Marketing* 64: 65–81.

Choi, Y., Y. Seo, and S. Yoon. 2017. E-WOM messaging on social media: social ties, temporal distance, and message concreteness. *Internet Research* 27: 495–505.

Chu, S., and Y. Kim. 2011. Determinants of consumer engagement in electronic word-of-mouth (eWOM) in social networking sites. *International Journal of Advertising* 30: 47–75.

Chu, S., and Y. Sung. 2015. Using a consumer socialization framework to understand electronic word-of-mouth (eWOM) group membership among brand followers on Twitter. *Electronic Commerce Research and Applications* 14: 251–60.

Chu, S., H. Chen, and Y. Sung. 2016. Following brands on Twitter: An extension of theory of planned behavior. *International Journal of Advertising* 35: 421–37.

Colliander, J., M. Dahlen, and E. Modig. 2015. Twitter for two: Investigating the effects of dialogue with customers in social media. *International Journal of Advertising* 34: 181–94.

eMarketer. 2015. Are sponsored social posts the most effective marketing channel? Usage and effectiveness of social ads have also increased. https://www.emarketer.com/Article/Sponsored-Social-Posts-Most-Effective-Marketing-Channel/1013242.

Edelman, B., S. Jaffe, and S.D. Kominers. 2016. To Groupon or not to Groupon: The profitability of deep discounts. *Marketing Letters* 27: 39–53.

Elwalda, A., K. Lu, and M. Ali. 2016. Perceived derived attributes of online customer reviews. *Computers in Human Behavior* 56: 306–19.

Ford, G.T., D.B. Smith, and J.L. Swasy. 1990. Consumer skepticism of advertising claims: Testing hypotheses from economics of information. *Journal of Consumer Research* 16: 433–41.

Forehand, M.R., and S. Grier. 2003. When is honesty the best policy? The effect of stated company intent on consumer skepticism. *Journal of Consumer Psychology* 13: 349–56.

Friestad, M., and P. Wright. 1994. The persuasion knowledge model: How people cope with persuasion attempts. *Journal of Consumer Research* 21: 62–74.

FTC. 2013. *.com Disclosures: How to make effective disclosures in digital advertising.* https://www.ftc.gov

Fu, P., C. Wu., and Y. Cho. 2017. What makes users share content on Facebook? Compatibility among psychological incentive, social capital focus, and content type. *Computers in Human Behaviors* 67: 23–32.

Goldsmith, R., and D. Horowitz. 2006. Measuring motivations for online opinion seeking. *Journal of Interactive Advertising* 6: 1–16.

Halliday, S.V. 2016. User-generated content about brands: Understanding its creators and consumers. *Journal of Business Research* 69: 137–44.

Hibbert, S., A. Smith, A. Davies, and F. Ireland. 2007. Guilt appeals: Persuasion knowledge and charitable giving. *Psychology & Marketing* 24: 723–42.

Holbrook, M.B., and R. Batra. 1987. Assessing the role of emotion as mediators of consumer responses to advertising. *Journal of Consumer Research* 14: 404–20.

Huang, A.H., K. Chen, D.C. Yen, and T.P. Tran. 2015. A study of factors that contribute to online review helpfulness. *Computers in Human Behavior* 48: 17–27.

Hwang, Y., and S. Jeong. 2016. "This is a sponsored blog post, but all opinions are my own": The effects of sponsorship disclosure on responses to sponsored blog posts. *Computers in Human Behavior* 62: 528–35.

IZEA. 2015. 2015 consumer state of sponsored social: A first wave baseline national study. https://izea.com/2015/11/03/consumers-rate-sponsored-social-as-equally-or-more-effective-than-all-other-forms-of-traditional-and-emerging-media/.

Jin, A.A., and J. Phua. 2014. Following celebrities' tweets about brands: The impact of twitter-based electronic word-of-mouth on consumers' source credibility perception, buying intention, and social identification with celebrities. *Journal of Advertising* 43: 181–95.

Jeong, H., and D. Koo. 2015. Combined effects of valence and attributes of e-WOM on consumer judgement for message and product: The moderating effect of brand community type. *Internet Research* 25: 2–29.

Kelley, H.H. 1973. The processes of causal attribution. *American Psychologist* 28: 107–28.

Kim, E., Y. Sung, and H. Kang. 2014. Brand followers' retweeting behavior on Twitter: How brand relationships influence brand electronic word-of-mouth. *Computers in Human Behavior* 37: 18–25.

Kim, A.J., and K.P. Johnson. 2016. Power of consumers using social media: Examining the influences of brand-related user-generated content on Facebook. *Computers in Human Behavior* 58: 98–108.

Kim, M., and M. Lee. 2015. Effects of review characteristics and consumer regulatory focus on perceived usefulness. *Social Behavior and Personality* 43: 1319–34.

Kozinets, R.V., K. de Valck, A.C. Wojnicki, and S.J.S. Winer. 2010. Networked narratives: Understanding word-of-mouth marketing in online communities. *Journal of Marketing* 74: 71–89.

Lee, H., Y. Han, K. K. Kim, and Y. Kim. 2014. Sports and social media: Twitter usage patterns during the 2013 super bowl broadcast. Proceedings of the International Conference on Communication, Media, Technology and Design, Istanbul, 2014. http://www.cmdconf.net/2014/pdf/40.pdf

Lee, J., D.H. Park, and I. Han. 2008. The effect of negative online consumer reviews on product attitude: An information processing view. *Electronic Commerce Research and Applications* 7: 341–52.

Lee, M., M. Kim, and W. Peng. 2013. Consumer reviews: Reviewer avatar facial expression and review valence. *Internet Research* 23: 116–32.

Lu, L., W. Chang, and H. Chang. 2014. Consumer attitudes toward blogger's sponsored recommendations and purchase intention: The effect of sponsorship type, product type, and brand awareness. *Computers in Human Behavior* 34: 258–66.

Lunardo, R., and A. Mbengue. 2013. When atmospherics lead to inferences of manipulative intent: Its effects on trust and attitude. *Journal of Business Research* 66: 823–30.

Obermiller, C., and E.R. Spangenberg. 2000. On the origin and distinctness of skepticism toward advertising. *Marketing Letters* 11: 311–22.

Paek, H., T. Hove, H.J. Jeong, and M. Kim. 2011. Peer or expert? The persuasive impact of YouTube public service announcement producers. *International Journal of Advertising* 30: 161–88.

Preacher, K.J., D.D. Rucker, and A.F. Hayes. 2007. Addressing moderated mediation hypotheses: Theory, methods, and prescriptions. *Multivariate Behavioral Research* 42: 185–227.

Reeder, G.D. 2009. Mindreading: Judgments about intentionality and motives in dispositional inference. *Psychological Inquiry* 20: 1–18.

Schultz, D.E., and J. Peltier. 2013. Social media's slippery slope: Challenges, opportunities and future research directions. *Journal of Research in Interactive Marketing* 7: 86–99.

Sung, Y., Y. Kim, O. Kwon, and J. Moon. 2010. An explorative study of Korean consumer participation in virtual brand communities in social network sites. *Journal of Global Marketing* 23: 430–45.

Till, B.D., and M. Busler. 2000. The match-up hypothesis: Physical attractiveness, expertise, and the role of fit on brand attitude, purchase intent and brand beliefs. *Journal of Advertising* 29: 1–13.

Tuk, M.A., P.W.J. Verlegh, A. Smidts, and D.H.J. Wigboldus. 2009. Sales and sincerity: The role of relational framing in word-of-mouth marketing. *Journal of Consumer Psychology* 19: 38–47.

Vermeir, I., and P. Van Kenhove. 2005. The influence of need for closure and perceived time pressure on search effort for price and promotional information in a grocery shopping context. *Psychology & Marketing* 22: 71–95.

Walther, J.B., B. Van Der Heide, L.M. Hamel, and H.C. Shulman. 2009. Self-generated versus other-generated statements and impressions in computer-mediated communication: A test of warranting theory using Facebook. *Communication Research* 36: 229–53.

Wei, M., E. Fischer, and K.J. Main. 2008. An examination of the effects of activating persuasion knowledge on consumer response to brands engaging in covert marketing. *Journal of Public Policy & Marketing* 27: 34–44.

Wei, P.S., and H.P. Lu. 2013. An examination of the celebrity endorsements and online customer reviews influence female consumers' shopping behavior. *Computers in Human Behavior* 29: 193–201.

Wood, N., and J.N. Burkhalter. 2014. Tweet this, not that: A comparison between brand promotions in microblogging environments using celebrity and company-generated tweets. *Journal of Marketing Communication* 20: 129–46.

Zaichkowsky, J.L. 1994. The personal involvement inventory: Reduction, revision, and application to advertising. *Journal of Advertising* 23: 59–70.

What does the brand say? Effects of brand feedback to negative eWOM on brand trust and purchase intentions

Manu Bhandari and Shelly Rodgers

ABSTRACT
This study examined the effects of brand feedback to negative electronic word of mouth (eWOM) on consumers' brand trust and purchase intentions, and examined the moderating effect of problem attribution described in the negative eWOM message. Results from a 2 (Brand Feedback: Present/Absent) × 2 (Causal Attribution: Brand/Reviewer) × 2 (Products: Laptop/TV) between-subjects online experiment suggest that brand feedback had a simultaneous positive and negative effect on purchase intentions, whereby brand trust mediated the positive effect. Attribution of the product's problem did not significantly change this effect. Findings help to clarify the literature by describing mechanisms through which brand feedback occurs to influence brand outcomes.

Introduction

Nielsen's (2015) global consumer survey showed that 66% of consumers trust other consumers' opinions in the form of online consumer reviews, or electronic word of mouth (eWOM). Scholarly findings show that eWOM can influence consumers' perceptions and behaviours online – including their brand trust and purchase intentions (e.g. Cheung and Thadani 2012; Sparks and Browning 2011).

Valence is a key concept used in eWOM studies (King, Racherla, and Bush 2014). Negative valence, defined as unfavourable feelings expressed toward the product or brand, leads to negative brand evaluations and lower purchase intentions (e.g. Cheung and Thadani 2012; Lee, Rodgers, and Kim 2009; Lee and Youn 2009; Sparks and Browning 2011). Much of the studies examining the effects of valence have examined consumer processing and effects of negative valence. However, too much focus on consumers' processing/ effects of negative eWOM without consideration of other factors that may influence information processing/effects of negative eWOM limits our understanding of eWOM.

Several scholars have begun investigating factors that may moderate negative eWOM effects, including consistency with other reviews (Kim and Gupta 2012; Quaschning, Pandelaere, and Vermeir 2015; Sparks and Browning 2011), regulatory focus (Zhang, Craciun, and Shin 2010), and product type (Sen and Lerman 2007). However, we may gain a fuller understanding of negative eWOM effects by expanding the focus from facets of eWOM under consumers' control versus those that are under the brands' or businesses' control (see Rodgers and Thorson 2000; Bhandari and Rodgers 2016). Brands and advertisers today often participate in the eWOM communication process by responding to negative product reviews by offering, for example, apologies or explanations to try and mollify an angry customer or to try and fix a problem with a product. The current study calls this 'brand feedback,' which refers to a company's written response to consumer feedback in eWOM setting that seeks to strengthen or uphold the company's promise to satisfy the consumers' needs and desires. The trend for businesses to provide brand feedback to customer complaints online has increased substantially in recent years (TripAdvisor 2013; Sparks and Bradley 2017). Even major online shopping sites like Amazon.com and Bestbuy.com allow brands/sellers to provide feedback to consumers. Yet, scholarship has not adequately kept pace to understand the role of the brand in the eWOM process. We lack a thorough knowledge of the effects brand participation would pose brand trust, purchase intentions, and other important variables. Although recommendations have been provided in both the academic (Sparks and Bradley 2017) and industry (TripAdvisor 2014) literature, such recommendations have not undergone thorough academic testing. Since both big and small brands continue to provide brand feedback to negative eWOM on online shopping sites, the lack of thoroughly tested recommendations for brands to strategize brand feedback creates an urgent need for a study to better understand the effects of brand feedback. Viewing the eWOM more dynamically – not as a one-way process – to also be a platform for brands to persuade inaccuracies or negative reviews – opens up a whole new area of research for eWOM.

Moreover, the effect of brand feedback is likely to interact with eWOM content attributes and persuasiveness, particularly attribution of the cause of the problem. Attribution means ascribing causality of the outcome described in the eWOM to a particular entity – for instance, the brand/product or the consumer writing the eWOM (i.e. the product reviewer). For example, if an eWOM message mentions poor customer service as a cause of the problem, consumers reading the eWOM probably will attribute the problem to the brand's failure. But when the eWOM message mentions a problem related to the online reviewer's characteristics (e.g. unfamiliarity with using the products), the reviewer is seen as the cause of the negative situation. In cases when brands or products are blamed less, negative eWOM may be less effective (Kim and Gupta 2012). These different causality attributions could produce different contexts within which consumers perceive and process brand feedback, and attribution of the cause could subsequently moderate the effects of brand feedback.

The purpose of this study thus was to conduct a 2 (Brand Feedback: Present/Absent) × 2 (Attribution: Brand/Reviewer) × 2 (Products: Laptop/TV) between-subjects experiment to examine the effects of brand feedback and attribution of the negative cause on consumers' brand trust and purchase intention in an online retail shopping context.

Literature review

Electronic word of mouth

eWOM is one of several sources of information consumers use to learn about the quality of products and services. Consumers who write eWOM or online product reviews (terms used interchangeably here) are believed to be candid about their views due to a perceived lack of financial motivation to assist the brand. Consumers may be influenced by eWOM messages because the messages have higher credibility, relevance, and ability to generate empathy (Bickert and Schindler 2001). Referring to eWOM allows consumers to gain confidence in their understanding about products, which consequently allows them to reduce their risk of making a bad purchase (Bronner and Hoog 2011; Cheng and Loi 2014; Goldsmith and Horowitz 2006; Hennig-Thurau and Walsh 2003). A major reason consumers use eWOM is to reduce the risk associated with making a bad purchase (e.g. Goldsmith and Horowitz 2006; Hennig-Thurau and Walsh 2003). This risk is heightened when eWOM spreads negative information about a brand or when consumers express dissatisfaction with some aspect of the brand or product (e.g. Blodgett, Hill, and Tax 1997; Moon, Costello, and Koo 2017). This has important implications for what consumers may look for in eWOM messages. If consumers want to reduce the risk of making a bad purchase, in the process of consuming eWOM they are likely to be attentive to cues of risks in buying the product under consideration.

Theoretically, negative eWOM provides risk cues that serve as warnings for consumers to avoid using a particular product. For example, a review for a skateboard that says 'a couple of the screws were loose' signals to other consumers to be wary of the product because it may be unsafe or not well made. Risk cues, in the form of negative reviews or eWOM, serve a diagnostic purpose helping consumers to categorize whether a product is lower quality (Lee and Youn 2009). Research continually demonstrates that negative eWOM negatively impacts brand trust, attitudes, and purchase intentions (e.g. Cheung and Thadani 2012; Lee et al. 2009; Lee and Youn 2009; Mauri and Minazzi 2013; Sparks and Browning 2011). And because negative cues (and the risk associated with making a bad purchase) are more salient than positive cues, negative eWOM is often more persuasive than positive eWOM (e.g. Lee, Rodgers, and Kim 2009; Lee and Youn 2009; Park and Lee 2009). This is why consumers pay attention to negative information conveyed in eWOM to lower or avoid purchase risks (Schindler and Bickart 2005). If negative eWOM has negative effects on brand trust and purchase intentions, brands should work to minimize such an effect.

Although negative eWOM helps consumers by reducing risk, research needs account for how other things may influence the process. In this regard, a potentially important trend has emerged in some eWOM environments: brands are becoming a part of the eWOM exchange of information. Given this scenario, we take a step further in this study by broadening the eWOM process to include brand participation. Brands can take steps to fend-off these negative effects by doing any number of things, e.g. a brand may empathize with a consumer who is disappointed with a product; a brand may offer to refund the consumer's money; a brand may offer to exchange the product for a new one; or a brand may offer accurate information to correct instances when consumers provide inaccurate details. No longer are brands at the mercy of consumers' negative reviews but they, too, can play a role. However, so far it is unclear how this brand tactic, i.e. responding to a negative comment, works. Below, we discuss the concept of brand feedback, and then we theorize about the effects of brand feedback.

Brand feedback

The word 'brand' may convey a variety of ideas (e.g. ownership of a product, monetary value, etc.). However, one important aspect is a brand's implied promise and trustworthiness to deliver something of value to consumers (Krugman and Hayes 2012). It is well known that 'brands make promises and seek to create trust between the producer and consumer' (Krugman and Hayes 2012, 439). Establishing trust is important for any firm or brand selling its products or services online (Cheng and Loi 2014; Krugman and Hayes 2012; Sparks and Browning 2011). Brand feedback, as conceptualized here, represents a brand's attempt to reinforce the validity of the brand promise and reinstate potential lost trust resulting from negative eWOM.

The literature on customer complaint handling suggests that brands need to respond properly to consumer complaints, as their responses can impact consumers' decision to be loyal to or spread negative word of mouth about the brand (Blodgett, Hill, and Tax 1997; Gelbrich and Roschk 2011). Traditionally, solving consumer complaints was relatively straightforward: communication took place between the complaining consumer and the brand, and the exposure of the interaction was more or less limited to the communicating parties only (Xie, Zhang, and Zhang 2014). However, the growth of the Internet and Web 2.0 has radically changed this. Now, consumers can simultaneously interact with and communicate between themselves *and* with brands on the same online platform. This presents a unique challenge to brands looking to advertise and promote their products and services online: the response to complaining consumers (i.e. negative eWOM contributors or reviewers) is now also available for other potential consumers to read, which can influence others' perceptions and buying-related decisions (e.g. Gu and Ye 2014; Sparks and Bradley 2017; Ye, Gu, and Chen 2010). Given the communication 'spread' that can happen rapidly online and with online social networks (Thorson and Rodgers 2017), there is a need for brands to respond to negative eWOM using a robust response strategy (e.g. Sparks and Bradley 2017).

Recent scholarship in the hospitality management literature has examined the possible influences of management or hotel managers' response, concepts very similar to brand feedback, to eWOM messages – specifically online reviews – on consumers' psychological and behavioural responses or business performance (Cheng and Loi 2014; Gu and Ye 2014; Litvin and Hoffman 2012; Mauri and Minazzi 2013; Sparks and Bradley 2017; Xie, Zhang, and Zhang 2014; Ye, Gu, and Chen 2010). Management response is defined as 'a business's effort to interact with and respond to consumer comments on experiences with the business or its products and services' (Gu and Ye 2014, 570). Both terms – brand feedback and management (or hotel managers') response – in an online context refer to a business responding to consumer feedback concerning the consumers' dissatisfaction with a brand.

Although the hospitality management literature has examined the effects of management response, no conclusive evidence about the effectiveness of feedback to negative eWOM has emerged. The mechanisms underlying brand feedback's effects are not yet fully understood. For instance, although some scholars have found that feedback to negative eWOM had a positive effect on sales, and improved attitudes toward purchase decision, and brand perceptions and brand reputations (e.g. Litvin and Hoffman 2012; Sparks and Bradley 2017; Ye, Gu, and Chen 2010), other studies have found that brand feedback reduces purchase intentions (Mauri and Minazzi 2013), or have mixed results (Cheng and Loi 2014; Xie, Zhang, and Zhang 2014).

One reason for inconsistent findings could be the existence of two separate bidirectional mechanisms through which brand feedback affects purchase intentions. In short, the following lines present a synopsis of the two (bidirectional) mechanisms or routes proposed and tested in the study. Mechanism 1: On one hand, the presence of brand feedback could send a positive message that the brand cares, or at least cared enough to respond; stands by its promise; can be trusted; and wants to earn consumer loyalty, which increases brand trust and hence purchase intentions. Mechanism 2: On the other hand, brand feedback could also send negative cues concerning post-purchase hassles of returning the product, including dealing with customer service. Next, we also know customers that can sometimes get suspicious about brand quality in an eWOM context when things seem unusual (e.g. too high or only positive ratings; see, for instance, Maslowska, Malthouse, and Bernritter 2017; Schindler and Bickert 2005). In a similar sense, seeing a brand feedback could raise consumers' suspicions, i.e. 'Why did the brand do that?' or 'Are they just trying to make me think they care about their consumers?' For example, while a consumer might appreciate that the brand took the time to care and respond to a negative review, logically, the brand wants consumers reading the review to buy more products from them. So brands may be expected to tell consumers anything to reach their goal. Because of all these potential negative cues, brand feedback may heuristically reduce purchase intentions independent of the positive effect via brand trust.

Below, we expand on the two proposed mechanisms.

Mechanism 1: positive effects of brand feedback

Some studies indicate that brand feedback may have positive influences on consumers reading negative eWOM (Litvin and Hoffman 2012; Sparks and Bradley 2017; Ye, Gu, and Chen 2010). Positive impact may occur when brands give strong arguments in their online feedback (Cheng and Loi 2014) or when customers' complaints are about aspects that the business cannot easily change or improve (Xie, Zhang, and Zhang 2014). Further, dialogic communication with consumers has also been found to be effective on consumer outcomes in social media context (Colliander, Dahlén, and Modig 2015). The upcoming paragraphs will zero in on brand feedback's potential positive effects on the variables of most concern to the study: brand trust and purchase intentions.

Based on Gefen's (2000) definition of trust, brand trust here refers to consumer's confidence that a brand will behave favourably or as expected. According to Wang and Emurian (2005), trust is a complex and abstract concept but usually involves four characteristics: a party that trusts and another that is trusted (e.g. consumers who trust and business/brands that are trusted); uncertain/risky situations that produce a need for trust (e.g. risky online credit card transactions, unknown product quality); leads to actions such as risk-taking behaviours (e.g. making a purchase online); and subjectivity (e.g. consumers have different degree of trust needs to purchase something). Trust is important to examine in studies about online purchases (Wang and Emurian 2005) and the positive influence of trust on purchase intentions has been established in the literature (Chang, Cheung, and Lai 2005; Cheng and Loi 2014; Sichtmann 2007). Brand trust may come from perceptions of brand credibility and competence, which leads to higher purchase intentions (Sichtmann 2007), but brand trust suffers when consumers harbour negative emotions about brands (Moon, Costello, and Koo 2017).

Trust has also been studied in the context of eWOM effects (e.g. Cantallops and Salvi 2014; Sparks and Browning 2011). Previous experiences with a brand allow consumers to compare the brand-supplied information and the actual quality from their own personal experience, and this can impact credibility perceptions, which in turn can enhance trust (Sichtmann 2007). Potential consumers who wish to know whether a brand or product will behave favourably may refer to the online product review, which provides information about other consumers' experiences with the product (Sparks and Browning 2011). In such cases, reading other consumers' negative or poor brand/product experiences communicated in negative online product reviews will negatively impact the potential consumers. If consumers feel negative emotions, which impact brand trust (Moon, Costello, and Koo 2017), upon reading the eWOM message, brands can seek to reinstate or build trust by showing through brand feedback that it cares about its customers' satisfaction. Caring about customers is an important component of trust (Gefen, Karahanna, and Straub 2003). Perceived caring has also been found to partially mediate brands' dialogic communication's effects on brand attitudes and purchase intentions in the context of social media platforms like Twitter (Colliander, Dahlén, and Modig 2015). Also, a survey showed that 77% of respondents reported that seeing a hotel management respond to online reviews made them feel that the hotel cared more about its guests (TripAdvisor 2014). This industry finding suggests that consumers may pay attention to brand feedback, and this may have implications for their perceptions of how much a brand cares about its consumers. Brand feedback can also show the integrity of brands to make good on their implied promise with their products, and their benevolence to take care of the consumer interest before profit. Both of these qualities – integrity and benevolence – are among the core dimensions of trust according to previous conceptualizations of trust in the literature (see Gefen et al. 2003).

From the above discussion, we posited that when negative eWOM would decrease consumers' trust in the business or brand, brand feedback should reinstate some of that trust back. Litvin and Hoffman (2012) found hotel's response to negative reviews had a positive effect on brand attitudes, although the effect was lower than when other consumers posted a positive response to a negative review. Similarly, through a comprehensive review of interdisciplinary literature on online reviews, Sparks and Bradley (2017) concluded that brand feedback would be better for improving a brand's reputation or image rather than no brand feedback at all.

Studies have also shown that brands that demonstrate care for their consumers, including solving any post-purchase problem, influences loyalty in an online shopping context (Srinivasan, Anderson, and Ponnavolu 2002). We have argued that brand feedback is the brand's attempt to show consumers that the brand cares and is worthy of their continued trust and loyalty. Hence, by creating or reinstating brand trust via brand feedback in eWOM, brand feedback becomes a means to reduce risk perception created by the negative eWOM about the featured product. Also, when brands communicate in a dialogic manner (rather than just sending information to consumers in a one-way manner), improves brand attitudes and purchase intentions on social media sites like Twitter, emphasizing the importance of engaging with customers (Colliander, Dahlén, and Modig 2015). Therefore, we posited that brand feedback should have a positive effect on purchase intentions via the mediation of brand trust:

H1: The presence (versus absence) of brand feedback to a negative review will lead to higher purchase intention ratings through an increase in (i.e. mediation of) brand trust.

Mechanism 2: possible negative effect of brand feedback

Brand feedback to negative eWOM has also yielded negative effects (Cheng and Loi 2014; Mauri and Minazzi 2013; Xie, Zhang, and Zhang 2014). For instance, management responses have lowered purchase intentions (Mauri and Minazzi 2013). As noted, a main reason consumers use eWOM messages is to avoid making a bad purchase (e.g. Goldsmith and Horowitz 2006; Hennig-Thurau and Walsh 2003). In the case of negative brand feedback, the very *act* of providing brand feedback may be a risk cue for consumers. This risk cue may in turn lead to lower intentions to purchase.

For example, past studies have shown that addressing consumer problems may not translate into loyalty to buy from the brand again (Holloway and Beatty 2003). Resolving a post-purchase problem can take time and may consist of multiple contacts with brands via different channels (Holloway and Beatty 2003). In fact, many consumers may not complain about brands since complaining takes time and effort (Holloway and Beatty 2003).

Thus, when consumers perceive any cue about burdensome post-purchase problems or interactions with customer service as a result of the brand feedback, it may trigger a mental schema that prompts consumers to avoid purchasing the product. Since consumers want to avoid unnecessary post-purchase interactions with customer service, the very presence of brand feedback could deter consumers from buying the online product. This issue is compounded by the fact that consumers may sometimes think that brand feedback is the businesses' attempt to promote itself, similar to paid advertising (Mauri and Minazzi 2013) or that the brand/business seeks to hide its mistakes through an eWOM response (Cheng and Loi 2014). Thus, even if brands provide brand feedback within the context of eWOM, the consumer would likely avoid the product since there are so many alternative choices available online. By acknowledging a problem with the product, brand feedback can alert other consumers that the product may be problematic.

Given the above discussion, the presence of brand feedback could independently exert a negative heuristic influence on people's intention to purchase a product, even though there is a simultaneous positive influence through an increase in brand trust. People may not just analyse the message addressing the problem but may use brand feedback to negative eWOM as a risk cue, indicating that the product could require post-purchase actions (i.e. returns). Hence, it was hypothesized that:

H2: The presence (versus absence) of brand feedback to a negative review will directly lower purchase intention ratings.

Attribution

Negative eWOM with different attributions may exert differing influences. For example, attribution of the cause of the problem is often made to the reviewer (i.e. eWOM contributor) or to the product/brand or to circumstances (Kim and Gupta 2012; Lee and Youn 2009; Quaschning, Pandelaere, and Vermeir 2015; Sen and Lerman 2007). When the problem described in a negative consumer eWOM is attributed to the reviewer's personal disposition, consumers tend to evaluate products more positively and judge eWOM to be less helpful, useful, or informative (Kim and Gupta 2012; Quaschning, Pandelaere, and Vermeir 2015; Sen and Lerman 2007) and have a poorer attitude toward the eWOM (Sen and Lerman 2007). Similarly, when a positive product review's cause is attributed to circumstances (e.g. a person's vested interest), rather than to the product's intrinsic positive quality, consumers judge the product more harshly (they are less likely to recommend the product to friends, Lee and Youn 2009). Given the preceding arguments, we hypothesized:

H3: Attribution of the product's problem to the brand will lower purchase intention ratings through a reduction in (mediation of) brand trust.

Additionally, brand feedback's influence may also differ based on attribution in the eWOM message. Negative eWOM that attributes product failures to the reviewer versus the brand should probably influence brand feedback's effects on trust and purchase intentions differently. Logically speaking, when consumers write reviews indicating that they are the potential cause of the problem with the product's performance (e.g. they cannot read instructions well, they are not sure if they assembled it correctly, etc.), other consumers who read the review will perceive brand feedback differently than when negative eWOM is attributed to the brand or product (e.g. product was broken during shipping, product did not work as indicated, etc.). To explore this, the following research question was posed:

RQ1: Will the presence of brand feedback lead to an increase in brand trust and purchase intentions for reviewer-attributed versus brand-attributed eWOM?

Method

The study used a 2 (Brand Feedback: Present/Absent) × 2 (Causal Attribution: Brand/Reviewer) × 2 (Products: Laptop/TV) between-subjects experimental design. The experiment was conducted on Qualtrics, a popular online research platform.

Measures

Independent variables
Brand feedback was operationalized as a written response by a business to consumers' reviews or eWOM message seeking to uphold the business' promise to satisfy the needs or desires of the consumers. The variable was manipulated as presence/absence of a brand feedback to a negative online review (see Table 1).

Attribution was the second independent variable but the study also considered its role as a moderator of brand feedback effects. Attribution was operationalized as the entity (either brand or reviewer) shown to be the cause of the situation as described in the negative product review. Attribution was manipulated as either attribution to the brand or to the reviewer (see Table 1).

Dependent and mediating variables
Brand trust was used as a dependent variable and a potential mediating variable. Brand trust was defined as consumers' beliefs that a brand can be depended on for favourable outcomes (Gefen 2000). Brand trust was measured on a 7-point Likert scales asking consumers to rate whether they felt the brand 'is honest'; 'can be trusted'; 'cares about its

Table 1. Electronic word of mouth (online product reviews) used in the study.

Product category	Brand attributed reviews	Reviewer attributed reviews	Brand feedback*
Dell laptop	The product arrived with some crack in it, but when I called Dell's customer service they just kept transferring me from this person to that person without really addressing my concerns for quite a while. Other message board posts mention the same thing. Looks like Dell's forgotten about customer service.	I honestly don't like this laptop, as it's hard for me to use. First of all, since I've not used any Dell products before. I found it hard to set up. Next, I accidentally dropped it once and it immediately showed a crack. The laptop has a low price, but it isn't good enough for me. I'm stuck with this equipment now.	We at Dell want to ensure that every product gives its customers 100% satisfaction. So we are sad to hear that the product did not meet your needs. Please contact us off-list at 1-800-XXX-XXXX to allow us to assist you with your concern. We want to do everything we can to make you feel valued, including exchanging the product or returning it for a full refund.
LG HDTV	This TV broke soon after a year — nothing but a white screen with a few vertical lines. The warranty period is not long enough to cover it, so customer service is quick to brush off any mention of its responsibility. Other people I talked to had similar experiences. Not a good TV.	Not recommending the product because of what I had to go through. I could not handle the setting up part very well, as I had not set up any TV before. I later realized I had messed up the wiring. So I got it fixed after talking with some people. This ruined the excitement I had when I ordered the TV.	We at LG want to ensure that every product gives its customers 100% satisfaction. So we are sad to hear that the product did not meet your needs. Please contact us off-list at 1-800-XXX-XXXX to allow us to assist you with your concern. We want to do everything we can to make you feel valued, including exchanging the product or returning it for a full refund.

*Electronic word of mouth (online product reviews) used in the study. The same 1-800 number was used in both the brand feedback posts to maintain consistency. The last seven numbers in the phone number are crossed because they were randomly chosen and may belong to someone else.

customers'; and 'would provide me with good service if I bought its products online' (Gefen et al. 2003; Hassanein and Head 2007). Purchase intention was operationalized as participants' likelihood of buying the featured product. The variable was measured using three items on 7-point semantic differential scales: unlikely/likely; impossible/possible; and improbable/probable (Schlosser, White, and Lloyd 2006).

Covariates

Participants' prior brand attitudes were measured and controlled for. Seven-point semantic-differential scales using bipolar adjectives indexed people's favourability, liking, goodness, and positivity (Holbrook and Batra 1987) toward the featured brands. Perceived realism was also measured, and was defined as how much the consumer believed he/she would likely encounter the review on online shopping sites such as Amazon.com. It was measured using a 7-point Likert scale to the question: 'This above customer product review looks similar to the kind of reviews I generally see on sites like Amazon.com.' Involvement or engagement with the product category was also measured using 7-point Likert scales ranging from (1) strongly disagree to (7) strongly agree using the items 'appealing,' 'interesting,' 'important,' and 'relevant.'

Stimulus materials

Online reviews were gathered from shopping sites (e.g. Amazon.com) and were edited for the purpose of the study. Stimulus materials were then pre-tested with a group of undergraduate students ($N = 21$) to select the most representative negative reviews for each level of attribution and products. Negative reviews were selected since they tend to have stronger effects than positive reviews (e.g. Lee, Rodgers, and Kim 2009) and brands are more concerned about negative reviews damaging their reputation.

In the pre-test, participants rated a series of reviews in terms of whether the brand or the customer caused the situation described in the review. The two questions used were measured using 7-point semantic-differential scale. The first question ranged from reviewer caused the situation to reviewer did not cause the situation. Similarly, the second question ranged from brand caused the situation to brand did not cause the situation. Additionally, the participants also rated on a 7-point Likert scale how much they believed: 'This review looks a lot like the reviews I've seen on sites like Amazon.com or Walmart.com'). Let us call that 'product review realism' for the study's purpose. The product review that performed the best considering all three items together was chosen for the main study. More specifically, the product reviews that performed – overall – the best to represent each specific condition and did not have too low product review realism was selected for each product (e.g. the review selected for 'brand-caused' or attributed condition had high brand-caused-the-problem rating and low customer-caused-the-situation ratings, and relatively not too low product review realism score).

Participants and procedure

Participants were 453 undergraduate students recruited from a large Midwestern university. A college student sample was considered appropriate for the study since, practically speaking, most college students (OnCampus Research 2013), and theoretically speaking, college students are an acceptable and useful sample to test theoretical multivariate relationships (Basil, Brown, and Bocarnea 2002). Six responses were deemed outliers or had missing data and were dropped from additional analyses. The final sample ($N = 447$, mean age $= 18.75$ years) was mainly female (71%, $n = 316$) and Caucasian (81.9%, $n = 366$). Other race/ethnicities in the study were Asian (7.8%, $n = 35$), African-American (4.9%, $n = 22$), Hispanic (2.9%, $n = 13$), Other (2%, $n = 9$), Native American ($n = 1$), and Pacific Islander ($n = 1$). Participants were randomly assigned to and received an emailed link to one of eight treatment conditions. Upon informed consent, participants read the online product reviews and completed the study measures including demographics. Participants received extra credit for their participation.

Results

There was a significant correlation between participants' prior brand attitudes and each of the two dependent variables, brand trust ($r = 0.38$) and purchase intentions ($r = 0.45$), suggesting that prior brand attitudes is a necessary covariate. The study used Process (Hayes 2013), a regression-based mediation test procedure using bootstrapping technique, to simultaneously examine direct and indirect effects of each of the independent variables. When running indirect effect tests involving any of the two independent variables (i.e. either attribution or brand feedback), the effect of the other independent variable was controlled for.

The study ruled out product review realism as a control variable because of its low and non-significant correlation with the dependent variables, $r(447) < -0.1$. Similarly, initial analyses showed that product involvement also did not impact brand trust or purchase intentions above and beyond attitude toward brand. Hence, for parsimony, product review realism and product category involvement were dropped as covariates.

Skewness and kurtosis for the sole covariate (prior brand attitude) and dependent variables for each group were below or close to $+/-1$, indicating that data distribution was acceptably normal for analysis. There was a single cell with kurtosis value at 2.6, so some caution in data interpretation needs to be exercised. Box's M was not significant ($p > 0.05$) although Levene's test for equality of variance was significant for brand trust ($p = 0.018$).

A manipulation check was conducted that analysed participants' ratings on whether they felt (a) the brand or (b) the reviewer was the main cause of the situation in each online review they read. An independent-samples t-test showed that participants reading the reviewer-controlled reviews made causality attribution to the reviewer ($p < 0.001$), and participants reading the brand-attributed reviews made causality attribution to the brand ($p < 0.001$). The attribution manipulation was, therefore, deemed successful.

Hypothesis/RQ testing

H1 and H2 concerned the effects of brand feedback. H1 posited that the presence of brand feedback would lead to a higher intent to purchase the product via brand trust. Results showed that brand feedback presence had a significant indirect positive effect on purchase intention via brand trust, as a zero value did not lie between 5000-sample 95% bias corrected bootstrap confidence interval ($B = 0.31$, 95% CI $= 0.20$ to 0.44). Thus, H1 was supported.

H2 hypothesized that the presence of brand feedback would directly lower purchase intention. This direct negative effect was expected to be independent of above-mentioned positive effect on purchase intentions via brand trust. Results showed a significant direct negative effect of brand feedback on purchase intention, as there was no zero value between the confidence intervals ($B = -0.45$, SE $= 0.13$, $t = -3.4$, 95% CI $= -0.72$ to -0.19). Thus, H2 was supported.

H3 tested the indirect effects of product review attribution on purchase intentions through brand trust. Results showed that brand trust partially mediated the effect of attribution on purchase intentions. Zero did not appear between 95% bootstrap confidence intervals ($B = -0.40$, 95% CI $= -0.55$ to -0.28). As predicted, the higher the brand attribution, the more negative the indirect effect on purchase intention via brand trust. H3 was, therefore, supported. However, there also was a significant direct effect ($B = -0.60$, SE $= 0.14$, $t = -4.3$, CI $= -0.87$ to -00.32), which showed that attribution of a negative outcome to the brand also had a negative direct effect on purchase intentions, independent of the mediation of brand trust.

RQ1 asked whether attribution would moderate the effect of brand feedback on purchase intentions. To test the RQ, brand feedback was retained as the independent variable, attribution was the moderator, brand trust was the mediator, and purchase intention was the main dependent variable. Results showed no significant Brand Feedback $\times$ Attribution interaction effect.

Table 2. The product variable's moderation of attribution and brand feedback effects directly and indirectly via brand trust. Attribution was coded as 0 (reviewer) and 1 (brand); brand feedback was coded as 0 (absent) and 1 (present).

Product	Independent variable	Effect route	B (SE B)	t	95% CI
Dell laptop	Attribution	Direct	0.12 (0.19)	0.61	−0.25 to 0.48
		Indirect	−0.53 (0.10)		−0.75 to −0.35
	Brand feedback	Direct	−0.79 (0.18)	−4.45	−1.15 to −0.44
		Indirect	0.21 (0.07)		0.09 to 0.35
LG HDTV	Attribution	Direct	−1.25 (0.18)	−7.07	−1.61 to −0.91
		Indirect	−0.35 (0.07)		−0.5 to −0.22
	Brand feedback	Direct	−0.09 (0.19)	−0.49	−0.46 to 0.27
		Indirect	0.38 (0.08)		0.25 to 0.56

Note: Indirect effects were tested using bootstrapping estimation approach with 5,000 bootstrap sample. B = unstandardized effect size.

To explore for possible product-related differences, additional analyses were conducted using Process (Hayes 2013; see Table 2). Results showed that the negative effect of brand feedback on purchase intentions disappeared for the LG HDTV, although it remained for the Dell laptop. Brand feedback's positive effect through brand trust also had a higher magnitude for the LG HDTV than the Dell laptop. Results also showed the product moderated attribution's effects (see Table 2). Although the negative indirect effect of brand attribution through brand trust remained, it was significantly higher for Dell laptop than for LG HDTV. Brand attribution's negative direct effect that was significant in the previous analyses, however, no longer showed significance for Dell laptop. In other words, for Dell laptop, brand trust completely mediated the negative effect of brand attribution. For LG HDTV, brand trust continued to only partially mediate the effect.

Discussion

Brands today face a need to respond to customer complaints in the form of negative eWOM or online product reviews. In this study, we investigated the effects of brand feedback to negative eWOM on consumers' information processing of online product reviews. We also examined whether negative eWOM attributes, i.e. attribution, moderated the effects of brand feedback in this process. The study conceptualized eWOM communication as a dynamic process, allowing participation of consumers and brands, and this in itself is a meaningful contribution to the eWOM literature.

First, results showed that effects of brand feedback had an indirect positive effect (through an increase in brand trust) *and* a direct negative effect on consumers' purchase intentions suggesting that effects of brand feedback may not be as straightforward as prior research suggests. When brand feedback was present, brand trust and purchase intentions increased. This finding supports the argument that it is beneficial to provide brand feedback to negative eWOM messages (e.g. Sparks and Bradley 2017; Ye et al. 2010). Presumably, brand feedback helped to reinforce the brand's implied promise to deliver a product of value and thereby lowered the amount of risk cues that consumers 'took home' after reading a negative eWOM message. Therefore, conceptualizing eWOM as occurring in a dynamic environment is meaningful in that brands can become part of the eWOM conversation that may impact potential consumers.

However, the current results also indicated that brand feedback could simultaneously reduce purchase intentions, independent of the positive effect through brand trust. We theorized that brand feedback could trigger consumers' mental heuristics about post-purchase challenges (i.e. return issues or hassles), thereby leading to a lower intention to purchase the reviewed product. Given that there are so many choices available for products online, even the presence of a minor cue about potential post-purchase problems may prompt consumers to make alternative choices. Additionally, when brand feedback was present in negative eWOM, results suggested that brand feedback could serve as a cue that legitimized the product's problem. This inconsistency could lead consumers to perceive product problems to be more persistent or 'stable' (Folkes 1984; Weiner 1985), and thus more likely to occur again in the future, ultimately leading to lower purchase intentions.

Theoretically, the above findings help to clarify some vagueness in the literature about whether brand feedback in an online context leads to positive or negative effects. Studies in the past have found both positive and negative effects but have not been very clear about the mechanisms behind those effects. The current study helps to dispel confusion about whether brand feedback has positive or negative effects by showing that brand feedback may impact consumers' purchase intentions positively *and* negatively, simultaneously through two separate routes – one mediated by brand trust (positive) and another direct (negative). To some degree, the current study helps to show the complexity of brand feedback's effects that can bring both positive and negative outcomes.

The role of attribution was also explored with regard to negative eWOM attributed to the reviewer versus the product/brand. Prior research suggests that negative eWOM attributed to the reviewer is less harmful than those attributed to the product/brand (e.g. Kim and Gupta 2012; Quaschning, Pandelaere, and Vermeir 2015). Although attribution's effects on brand trust and purchase intentions were consistent with past studies, the current study, however, did not replicate the result when taking brand feedback into account. In other words, attribution of the product's problem did not significantly change the effect of brand feedback on the study's dependent variables.

Practically speaking, advertisers, marketers, and managers can use the current results in developing strategies about their brand feedback to consumer eWOM. Although scholars (e.g. Sparks and Bradley 2017) and industry players (TripAdvisor 2014) have offered some guidelines on brand feedback, the role of brand feedback in the context of eWOM has not been thoroughly tested in the academic literature. The current study showed that it is risky for brands to assume that brand feedback has positive brand benefits alone. Thus, while brand feedback did appear to be a useful way to enhance brand trust and purchase intentions, it also produced negative results that were detrimental to the businesses' interests. Although no single study can be expected to provide a complete picture, results do show that practitioners need to strategize their brand feedback to avoid or minimize potential negative outcomes. It may be useful to examine features of the negative eWOM content before responding. For instance, practitioners could check who gets blamed for product problems, whether brands or reviewers. Our results suggest that when online reviewers blame brands for product problems, brands should carefully consider if or how to respond to such complaints to avoid detrimental effects to brand trust and purchase intentions.

Limitations and future research

Future research should examine potential mediating variables causing the negative direct effects observed in this study. These potential mediating variables could include attitude toward products returning process, among several others. Additionally, the role of variables such as perceived justice, satisfaction, and/or negative emotion (Gelbrich and Roschk 2011; Moon, Costello, and Koo 2017) – potentially important mediators of brand feedback effects – also needs to be examined in future research. The use of an undergraduate student sample, although useful to test theoretical relationships between variables (Basil, Brown, and Bocarnea 2002), limits generalizability of the findings to the larger population. However, by trying to find causal relationships between variables and underlying principles the study contributes toward a better understanding of variables and their relationship with each other, which in turn can allow scholars to arrive at more meaningful generalizability (Shapiro 2002). Care should be used in selecting products for eWOM studies. Although a pre-test of stimuli confirmed the efficacy of stimulus materials, and several controls were used in relation to the products, results differed depending on the product (TV or laptop) viewed. Returning a television purchased online is much riskier and poses greater challenges for consumers than a laptop purchased online. Since the exact source of the observed product-related differential effects is unclear, future research should examine the potential moderators.

Conclusion

Overall, the study contributes to the growing eWOM literature. The study conceptualized eWOM as a dynamic process where brands can, indeed, hold influence on the eWOM conversation. The study's results provided support for this dynamic conceptualization, as the presence of brand feedback showed significant direct and/or indirect influence on consumers' purchase intentions. Hence, brands' participation does have implications for persuasion or promotion through eWOM, although these implications may be positive and/or negative. The current study clarifies some of the ambiguities regarding brand feedback's effects by showing that brand feedback can have positive and negative effects on purchase intentions simultaneously, at least for certain products. The study's findings show the need for brands to carefully strategize brand feedback in eWOM environments. Brands may want to avoid brand feedback's negative impact while still benefitting from its positive indirect impact on purchase intentions and brand trust, although more research is needed to shed light on all the major factors important to consider in brand feedback. Although attribution did not moderate brand feedback's effect, results do suggest that it would make more sense to focus brands' limited available time for brand feedback activities on brand-attributed problems. Brand feedback may be useful but not necessary for reviewer-attributed problems, or when brands are not directly blamed for the problems, as these types of problems have less negative impact on brands.

Disclosure statement

No potential conflict of interest was reported by the authors.

References

Basil, Michael D., William J. Brown, and Mihai C. Bocarnea. 2002. Differences in univariate values versus multivariate relationships. *Human Communication Research* 28, no. 4: 501–14.

Bhandari, Manu and Shelly Rodgers. 2016. Electronic word-of-mouth and user-generated content: Past, present and future. In *The new advertising: Branding, content, and consumer relationships in the data-driven social media era,* vol. 1, ed. Ruth E. Brown, Valerie K. Jones and Ming Wang, 175–201. Santa Barbara, CA: Praeger.

Bickart, Barbara and Robert M. Schindler. 2001. Internet forums as influential sources of consumer information. *Journal of Interactive Marketing* 15, no. 3:31–40.

Blodgett, Jeffrey G., Donna J. Hill, and Stephen S. Tax. 1997. The effects of distributive, procedural, and interactional justice on postcomplaint behavior. *Journal of Retailing* 73, no. 2: 185–210.

Bronner, Fred and Robert De Hoog. 2011. Vacationers and eWOM: Who posts, and why, where, and what?. *Journal of Travel Research* 50, no. 1:15–26.

Cantallops, Antoni Serra and Fabiana Salvi. 2014. New consumer behavior: A review of research on eWOM and hotels. *International Journal of Hospitality Management* 36: 41–51.

Chang, Man Kit, Waiman Cheung, and Vincent S. Lai. 2005. Literature derived reference models for the adoption of online shopping. *Information & Management* 42, no. 4: 543–59.

Cheng, Vincent T.P., and Mei K. Loi. 2014. Handling negative online customer reviews: The effects of elaboration likelihood model and distributive justice. *Journal of Travel & Tourism Marketing* 31, no. 1: 1–15.

Cheung, Christy M.K., and Dimple R. Thadani. 2012. The impact of electronic word-of-mouth communication: A literature analysis and integrative model. *Decision Support Systems* 54, no. 1: 461–70.

Colliander, Jonas, Micael Dahlén, and Erik Modig. 2015. Twitter for two: Investigating the effects of dialogue with customers in social media. *International Journal of Advertising* 34, no. 2: 181–94.

Folkes, Valerie S. 1984. Consumer reactions to product failure: An attributional approach. *Journal of Consumer Research* 10, no. 4: 398–409.

Gefen, David. 2000. E-commerce: The role of familiarity and trust. *The International Journal of Management Science* 28, no. 6: 725–37.

Gefen, David, Elena Karahanna, and Detmar W. Straub. 2003. Trust and TAM in online shopping: An integrated model. *MIS Quarterly* 27, no. 1:51–90.

Gelbrich, Katja, and Holger Roschk. 2011. A meta-analysis of organizational complaint handling and customer responses. *Journal of Service Research* 14, no. 1: 24–43.

Goldsmith, Ronald E., and David Horowitz. 2006. Measuring motivations for online opinion seeking. *Journal of Interactive Advertising* 6, no. 2: 2–14.

Gu, Bin, and Qiang Ye. 2014. First step in social media: Measuring the influence of online management responses on customer satisfaction. *Production and Operations Management* 23, no. 4: 570–82.

Hassanein, Khaled, and Milena Head. 2007. Manipulating perceived social presence through the web interface and its impact on attitude towards online shopping. *International Journal of Human-Computer Studies* 65, no. 8:689–708.

Hayes, Andrew F. 2013. *Introduction to mediation, moderation, and conditional process analysis: A regression-based approach.* New York: Guilford Press.

Hennig-Thurau, Thorsten, and Gianfranco Walsh. 2003. Electronic word-of-mouth: Motives for and consequences of reading customer articulations on the Internet. *International Journal of Electronic Commerce* 8, no. 2: 51–74.

Holbrook, Morris B., and Rajeev Batra. 1987. Assessing the role of emotions as mediators of consumer responses to advertising. *Journal of Consumer Research* 14, no. 3: 404–20.

Holloway, Betsy B., and Sharon E. Beatty. 2003. Service failure in online retailing: A recovery opportunity. *Journal of Service Research* 6, no. 1: 92–105.

Kim, Junyong, and Pranjal Gupta. 2012. Emotional expressions in online user reviews: How they influence consumers' product evaluations. *Journal of Business Research* 65, no. 7: 985–92.

King, Robert A., Pradeep Racherla, and Victoria D. Bush. 2014. What we know and don't know about online word-of-mouth: A review and synthesis of the literature. *Journal of Interactive Marketing* 28, no. 3: 167–83.

Krugman, Dean M., and Jameson L. Hayes. 2012. Brand concepts and advertising. In *Advertising theory*, ed. Shelly Rodgers and Esther Thorson, 434–448. New York: Routledge.

Lee, Mira, Shelly Rodgers, and Mikyoung Kim. 2009. Effects of valence and extremity of eWOM on attitude toward the brand and website. *Journal of Current Issues & Research in Advertising* 31, no. 2: 1–11.

Lee, Mira, and Seounmi Youn. 2009. Electronic word of mouth (eWOM): How eWOM platforms influence consumer product judgement. *International Journal of Advertising* 28, no. 3: 473–99.

Litvin, Stephen W., and Laura M. Hoffman. 2012. Responses to consumer-generated media in the hospitality marketplace: An empirical study. *Journal of Vacation Marketing* 18, no. 2: 135–45.

Maslowska, Ewa, Edward Malthouse, and Stefan F. Bernritter. 2017. Too good to be true: the role of online reviews' features in probability to buy. *International Journal of Advertising* 36, no. 1: 142–63.

Mauri, Aurelio G., and Roberta Minazzi. 2013. Web reviews influence on expectations and purchasing intentions of hotel potential customers. *International Journal of Hospitality Management* 34: 99–107.

Moon, Sun-Jung, John P. Costello, and Dong-Mo Koo. 2017. The impact of consumer confusion from eco-labels on negative WOM, distrust, and dissatisfaction. *International Journal of Advertising* 36, no. 2: 246–71.

Nielsen. 2015. Global trust in advertising: Winning strategies for an evolving media landscape. https://www.nielsen.com/content/dam/nielsenglobal/apac/docs/reports/2015/nielsen-global-trust-in-advertising-report-september-2015.pdf

OnCampus Research. 2013. Students shop online, but so what? *OnCampus Research Newsletter.* http://www.nacs.org/email/html/IND-033-07-13/OCR_Newsletter_0913.html

Park, Cheol, and Thae Min Lee. 2009. Information direction, website reputation and eWOM effect: A moderating role of product type. *Journal of Business Research* 62, no. 1:61–7.

Quaschning, Simon, Mario, Pandelaere, and Iris Vermeir. 2015. When consistency matters: The effect of valence consistency on review helpfulness. *Journal of Computer-Mediated Communication* 20, no. 2: 136–52.

Rodgers, Shelly, and Esther Thorson. 2000. The interactive advertising model: How users perceive and process online ads. *Journal of Interactive Advertising* 1, no. 1: 41–60.

Schindler, Robert M., and Barbara Bickart. 2005. Published word of mouth: Referable, consumer-generated information on the Internet. In *Online consumer psychology: Understanding and influencing consumer behavior in the virtual world*, eds. Curtis Hauvgedt, Karen Machleit and Richard Yalch, 35–61. Mahwah, NJ: Lawrence Erlbaum Associates.

Schlosser, Ann E., Tiffany Barnett, White, and Susan M. Lloyd. 2006. Converting web site visitors into buyers: How web site investment increases consumer trusting beliefs and online purchase intentions. *Journal of Marketing* 70, no. 2: 133–48.

Sichtmann, Christina. 2007. An analysis of antecedents and consequences of trust in a corporate brand. *European Journal of Marketing* 41, no. 9/10: 999–1015.

Sen, Shahana, and Dawn Lerman. 2007. Why are you telling me this? An examination into negative consumer reviews on the web. *Journal of Interactive Marketing* 21, no. 4: 76–94.

Shapiro, Michael A. 2002. Generalizability in communication research. *Human Communication Research* 28, no. 4: 491–500.

Sparks, Beverley A., and Graham L. Bradley. 2017. A 'Triple A' typology of responding to negative consumer-generated online reviews. *Journal of Hospitality & Tourism Research* 41, no. 6:719–45.

Sparks, Beverley A., and Victoria Browning. 2011. The impact of online reviews on hotel booking intentions and perception of trust. *Tourism Management* 32, no. 6: 1310–23.

Srinivasan, Srini S., Rolph Anderson, and Kishore Ponnavolu. 2002. Customer loyalty in e-commerce: An exploration of its antecedents and consequences. *Journal of Retailing* 78, no. 1: 41–50.

Thorson, Esther, and Shelly Rodgers. 2017. Network advertising model (NAM). In *Digital advertising: Theory and research*. 3rd ed. eds. Shelly Rodgers and Esther Thorson, 19–30. New York: Routledge and Taylor & Francis Group.

TripAdvisor. 2013. Management response contributions more than double from 2011 to 2012. (Retrieved May 2015, currently unavailable on site). http://www.tripadvisor.com/TripAdvisor-Insights/n604/management-response-contributions-more-double-2011-2012

Trip Advisor. 2014. How to add management responses to tripadvisor traveler reviews. http://www.tripadvisor.com/TripAdvisorInsights/n2428/how-add-management-responses-tripadvisor-traveler-reviews

Wang, Ye D., and Henry H. Emurian. 2005. An overview of online trust: Concepts, elements, and implications. *Computers in Human Behavior* 21, no. 1: 105–25.

Weiner, Bernard. 1985. An attributional theory of achievement motivation and emotion. *Psychological Review* 92, no. 4: 548–73.

Xie, Karen L., Zili Zhang, and Ziqiong Zhang. 2014. The business value of online consumer reviews and management response to hotel performance. *International Journal of Hospitality Management* 43: 1–12.

Ye, Qiang, Bin Gu, and Wei Chen. 2010. Measuring the influence of managerial responses on subsequent online customer reviews – a natural experiment of two online travel agencies. https://doi.org/10.2139/ssrn.1639683

Zhang, Jason Q., Georgiana Craciun, and Dongwoo Shin. 2010. When does electronic word-of-mouth matter? A study of consumer product reviews. *Journal of Business Research* 63, no. 12:1336–41.

The interconnected role of strength of brand and interpersonal relationships and user comment valence on brand video sharing behaviour

Jameson L. Hayes, Yan Shan and Karen Whitehill King

ABSTRACT

This study addresses gaps in our understanding of viral advertising by examining the following questions: (1) how do brand relationships, interpersonal relationships, and user comment valence influence decisions to accept referrals of and re-share online video ads, and (2) how do the roles of each intertwine to influence those decisions. A national sample ($N = 405$) of 18- to 34-year-old Facebook users participated in a 2 (stronger/weaker brand relationship strength) × 2 (stronger/weaker interpersonal relationship strength) × 2 (positive/negative user comments) experiment. The results show that interpersonal relationships, brand relationships, and valence of user comments play different, but intertwining roles in social networking site users' acceptance of viral video ads and decisions to re-share. Interpersonal relationships most strongly influenced ad referral acceptance while brand relationships and comment valence drove re-sharing. A three-way interaction, however, revealed that positive user comments greatly strengthen the brand's impact on referral acceptance; negative comments strongly negate brand influence.

Introduction

The worldwide online video audience is projected to grow to 2.15 billion in 2017 (eMarketer 2016a). The surge of video consumption has led to an increase in US online video advertising expenditures to a projected $16.7 billion in 2020 with 70% of marketers expecting to invest in social video ads specifically by mid-2017 (eMarketer 2016b). The popularity of online video ads relies on the high degree of consumer engagement generated from videos' creativity, interactivity, and shareability. Online video ads increase audience engagement, garner higher click-through rates (Rosenkrans 2009), and generate higher brand recall than non-video advertising formats (Lee, Ham, and Kim 2013).

Online video ads have the potential to go viral, and advertisers are trying to capitalize on the social-sharing instinct among Internet users (eMarketer 2016b). Viral advertising is a new hybrid of traditional advertising and electronic word-of-mouth (eWOM) that can be defined as marketer-initiated unpaid peer-to-peer communication of branded content with persuasive intent (Cho, Huh, and Faber 2014; Porter and Golan 2006). According to Porter and Golan (2006), three characteristics make viral advertising different from traditional advertising (e.g. print ads, television commercials) and eWOM (e.g. product reviews, consumer testimonials). First, viral advertising campaigns have no paid media involved. They are typically seeded through existing consumers or brand followers' social connections, which allow advertising content to spread across the social networks organically, garnering earned media to leverage brand awareness. Second, while traditional advertising is non-personal (one-to-many), viral advertising is personal (many-to-many). A successful viral advertising campaign requires content to be distributed by trusted sources, mostly friends or family. Lastly, compared with other forms of eWOM communication in which the messages might come from anonymous sources, viral advertising is from an identified sponsor. Having two sources – the brand that crafts the message and the social networking site (SNS) user sharing it, viral video ads combine the consumer endorsement power of eWOM with the message control and high reach potential of traditional advertising.

Brands are jumping on the bandwagon of viral advertising. According to *Advertising Age*, Shell's animated clean energy campaign topped viral video views for 2016. The video 'Make the Future,' featuring six award-winning international musicians and six energy innovations, garnered 262 million views after launching in September (see Bhattacharyya 2016). Viral video advertising within SNSs is particularly attractive for brands because the platforms offer (1) interactive social networks for video dissemination and (2) platforms for consumer–brand relationship (CBR) development via engagement with consumers and their social networks (Kaplan and Haenlein 2010; Mangold and Faulds 2009). As such, brands will pour about $36 billion into social network ad spending in 2017 hoping to drive viral activity (eMarketer 2015).

Until recently, viral advertising research largely ignored the interplay between brands, consumers, and their interpersonal contacts focusing primarily on sharing motivations (e.g. Lee, Ham, and Kim 2013), message effects (e.g. Eckler and Bolls 2011), and platform effects (e.g. Chu 2011). Recently, Hayes and King (2014) provided the first empirical evidence that brand relationships influence decisions to post and pass along viral video ads. Further, Cho, Huh, and Faber's (2014) viral email study suggests that both brand and interpersonal factors influence viral email acceptance; however, it is unclear how these results translate to the SNS one-to-many communication environment.

Additionally, while the past research provides progress towards an understanding of the brand–consumer social network eWOM dynamic in viral advertising decisions, a vital aspect is still overlooked: the commentary users add when sharing. On Facebook, for example, a user can share a video ad adding their own comments, which represents a wild card for brands in that comments may reinforce the brand or content (e.g. 'I love this video, so funny') or be oppositional (e.g. 'The worst that I have ever seen'; Campbell et al. 2011). However, studies examining the role of comment valence in viral advertising are scarce. Previous studies on message valence were largely conducted in a product review context; the findings on its effects in shaping consumer decision-making generated mixed results. A number of studies have found that review valence influences product evaluation, suggesting that online reviews influence product evaluation in the same direction as the valence of online reviews (Lee, Park, and Han 2008; Zhu and Zhang 2010; Willemsen et al. 2011). Despite this, other studies have found no effects for review valence (Liu 2006; Duan, Gu, and Whinston 2008; Cheung et al. 2009). To date, no published research has examined the influence of comment valence on the acceptance and sharing of viral video advertising.

This study seeks to address gaps in knowledge by examining how these variables combine to influence viral video referral acceptance and re-sharing intentions. Following previous viral advertising research, the present research defines referral acceptance as the decision to click on and exposes one's self to a referred ad; re-share intention is the intention to pass along a referred ad to one's own social network (Hayes and King 2014; Hayes, King, and Ramirez 2016). Specifically, the following questions are investigated: (1) how do brand relationships, interpersonal relationships, and user comment valence influence decisions to accept referrals of and re-share online video ads? and (2) how do the roles of each intertwine to influence those decisions?

Brand content sharing as social exchange

Viral advertising involves brand-generated message dissemination through online interpersonal networks. Social exchange theory (SET) is a useful lens for examining viral decisions due to its application across three areas vital to understanding sharing of ads: (1) interpersonal relationships, (2) CBRs, and (3) content sharing (see also Hayes and King 2014).

SET is a social psychology theory with the seminal purpose of conceptualizing the construction and maintenance of interpersonal relationships. All social exchanges are viewed as cost–benefit propositions wherein actors assess the expected value of exchanges before proceeding (Frenzen and Nakamoto 1993). Benefits garnered in exchanges can be tangible (e.g. future quid pro quo) and intangible (e.g. reputation, self-worth; Blau 1964). Multiple positive reciprocal exchanges between partners lead to trust in, intrinsic value for, and emotional attachment to exchange partners. Trust then leads to commitment to the interpersonal relationship allowing for reduced uncertainty in exchanges and the securing of continued benefits (Molm 1990; Molm, Takahashi, and Peterson 2000; Lawler, Thye, and Yoon 2000).

For example, if Actor 1 posts content on Facebook that Actor 2 provides positive affirmation for via liking or commenting, Actor 1 may reciprocate providing a positive response to future Actor 2 posts, and so forth. A series of net positive and equitable exchanges between the two Facebook users will result in emotional attachment to the exchange partnership generating trust. Facebook friends also assign intrinsic value to one another using their interactions for self-expressive purposes and commit to continued interaction mitigating uncertainty and securing continued benefits (Lawler, Thye, and Yoon 2000; Molm, Takahashi, and Peterson 2000). Therefore, Facebook sharing will persist between the friends.

The CBR perspective adopts this SET conceptualization of interpersonal relationship development as a metaphor for brand relationship development (e.g. Fournier 1998). Brands are seen as active relationship partners creating interdependencies with consumers through engaging in a mutually beneficial exchange. Like interpersonal relationships, CBRs evolve through multiple satisfactory exchanges between the consumer and the brand; brand satisfaction is the antecedent to brand trust. Brand satisfaction and trust determine the level of commitment consumers have to the brand (Hess and Story 2005). Further, humans often use brands to express and define themselves and even anthropomorphize brands as seen in interpersonal relationships (Bourdieu 1984; Brown 1991). With social media, brands collaborate with consumers more than ever in the initiating and maintaining of relationships. As such, brand exchanges too are viewed as cost–benefit analyses wherein consumers look to maximize benefit and minimize uncertainty through committing to a trusted brand (Morgan and Hunt 1994). So, similar to the previous interpersonal relationship example, actors' relationships with the brand subject of the shared ad likely influence decisions as prior experiences with the brand have yielded levels of emotional attachment, trust, and expressive value ascribed to the brand.

Of course, the content being shared is also a key part of the viral sharing process. Constant, Kiesler, and Sproull (1994) draw upon SET in conceptualizing sharing online as exchange and expressive functions. Content sharing is pragmatic in that actors share to receive reciprocal benefits and be self-expressive. Sharing decisions are a function of a cost–benefit analysis. Hayes and King (2014) found evidence of this in the viral advertising context as ad referral decisions were driven by expected benefits, reputation, and expected relationships. However, the research focused only on brand-generated content – the viral ad – excluding the user comments that are potentially an important part of the sharing decision.

This study seeks to investigate the dynamic social processes of consumer engagement with viral video ads. In the viral advertising context, consumers are presented with a dual-sourced ad – the brand and the interpersonal contact. From an SET standpoint, decisions to accept ad referrals and to re-share ads involve cost–benefit analyses wherein the ad recipient relies upon their relationships with the brand and their interpersonal contacts (which are constructed through a history of interactions) in assessing the expected value of the potential exchange (viewing/re-sharing the ad). Further, user comments added by the interpersonal contacts provide additional context as to the value of ad content to factor into the decisions. We posit that SNS users' sharing decisions are influenced by the CBR, the interpersonal relationship between source and receiver, and the valence of user-generated comment towards the referred ad.

Interpersonal relationships and sharing decisions

Viral advertising and general eWOM literature consistently note personal contacts as integral to message effectiveness (e.g. Chu and Kim 2011). Messages delivered through trusted personal contacts are perceived as more credible and as having less persuasive intent than traditional advertising (van Noort, Antheunis, and van Reijmersdal 2012). This leads to a larger impact on consumer expectations (Anderson and Salisbury 2003; Harrison-Walker 2001), attitude towards the ad (Herr, Kardes, and Kim 1991), brand attitude (Godes and Mayzlin 2004), and purchase decisions (Richins 1983).

Trust is essential to the exchange and integration of information as well as judging value in peer-to-peer information exchanges (Jarvenpaa, Knoll, and Leidner 1998; Zeng and Reinartz 2003). SNS users generally maintain a wide array of interpersonal connections within their social networks ranging from family members and close personal contacts to acquaintances or strangers all of which can refer ads in the SNS environment. Tie strength – defined in terms of the degree of relationship closeness, intimacy, frequency of interactions, and reciprocity between sender and receiver (Granovetter 1973) – lends trust and perceived value to the brand message enhancing referral acceptance and passing (Chiu et al. 2007; Phelps et al. 2004).

Chu and Kim (2011) found that tie strength and trust were positively associated with eWOM behaviour in SNSs. Further, Cho, Huh, and Faber (2014) found that a viral ad email sent by a friend or family member, as compared to one sent by an unknown other, is more likely to be noticed and opened, as well as generating positive attitudes towards the advertised brand. Chiu et al. (2007) found that email messages from close friends were more likely to be passed along than those from acquaintances. Therefore, interpersonal relationships will likely play an active role in SNS users' evaluation of viral video ads and decisions to re-share.

H1: The relationship strength between the sender and the recipient has a positive influence on the acceptance of a viral video ad (referral acceptance – RA).
H2: The relationship strength between the sender and the recipient has a positive influence on the intention of sharing a viral video ad (re-share intention – RI).

Brand relationships and sharing decisions

Understanding brand influence is essential in studying viral advertising because the brand is the original message source (Nan and Faber 2004). In the SNS environment, brands are more actively building relationships with their consumers using channels such as Facebook and Twitter to frequently communicate with SNS users. Consumers can interact with a brand by liking or sharing a brand message, becoming fans of a brand page, commenting on a brand status, following a brand, and creating creative brand-related content. The interactions between consumers and brands on SNSs are more likely to foster positive exchange relations (Mangold and Faulds 2009). Through these interactions, brand trust and commitment can be developed. Fournier and Alvarez (2012) suggest that the CBR combines cognitive, emotional, and behavioural processes in which trust and commitment are developed between brands and consumers. Therefore, the relationship with a particular brand is not only functional as consumers obtain the utilitarian benefits of the brand, but also expressive, providing consumers with important symbolic and social meanings. Hayes and King (2014) illustrate that trusted brands are leveraged as conduits for social benefit acquisition in viral advertising sharing as stronger CBRs reduce perceived uncertainty, enhance positive evaluation, and encourage message acceptance (see also Chu 2011; Hayes 2016; Shan and King 2015). Thus,

H3: The relationship strength between the brand and the recipient has a positive influence on referral acceptance.
H4: The relationship strength between the brand and the recipient has a positive influence on re-share intention.

User comment valence and sharing decisions

In traditional settings, marketers can control advertising information, respond to negative publicity about a brand and attempt to contain it by using various marketing strategies. However, user-generated content (UGC) on the Internet, as a form of eWOM, complicates this issue by shifting the control of marketing information from marketer to consumer. Consumers can say what they want online, which inevitably contains negative and positive details about a particular product or brand. In the context of viral advertising, the dissemination of advertising content relies on SNS users' sharing and pass-long behaviour and allows co-creation by inviting consumers to comment on the shared content. Thus, it is important to understand how positive and negative UGCs impact consumer evaluation and decision-making processes.

According to Fishbein and Ajzen (1975), message valence refers to audiences' subjective evaluation of the consequence associated with the target belief. Message valence has been conceptualized by a bipolar attitude model on a one-dimensional measurement with positive and negative inclinations at two extremes, and neutral evaluation in the middle (Kaplan 1972). A review of the literature on UGC shows that levels of message valence play varying roles in influencing consumers' information processing and product evaluations (Tang, Fang, and Wang 2014). Studies have found that positive consumer reviews have a positive impact on product sales as eWOM offers additional information about product attributes and performances that are independent from marketers' influences (Berger, Sorensen, and Rasmussen 2010; Liu 2006; Chevalier and Mayzlin 2006). In contrast, negative consumer reviews about a brand are more likely to influence brand evaluations and product judgement as negative information, which is viewed as more diagnostic, tends to be weighted more heavily than positive information (Lee and Youn 2009; Lee, Rodgers, and Kim 2009).

However, there is a scarcity of studies examining how the valence of user comments influences users' attitudes and sharing intentions towards a viral video ad on SNSs. According to SET, motivation is a key determinant of social exchange behaviour; exchanges occur when individuals are motivated by altruism, expectation of reciprocal benefits, self-expression, and self-enhancement (Smith et al. 2007; Hsu and Lin 2008; Taylor, Strutton, and Thompson 2012). Previous literature suggests that consumer attitude towards the advertisement and the likelihood of sharing an ad are influenced by reciprocal altruism and expected positive outcomes (Hayes, King, and Ramirez 2016; Kiyonari, Tanida, and Yamagishi 2000). Rim and Song (2016) found that positive comments enhance sharing motivation subsequently eliciting favourable brand attitudes in corporate social responsibility campaigns; examining user comments in consumer-run brand community sites, Jeong and Koo (2015) report that negative subjective comments evoked strong negative attitudes while positive subjective comments lead to less strong positive attitudes. Therefore, positive user comments regarding the referred viral video ad are expected to create higher levels of motivation leading to expectations of positive outcomes of consuming and sharing the viral video ad; the opposite is expected for negative user comments.

H5: The valence of user comments has a positive relationship with referral acceptance.
H6: The valence of user comments has a positive relationship with re-share intention.

Recent studies have reported that brand influence and interpersonal influence can supplement one another to impact consumer evaluation and sharing intention of a viral advertising message. For example, Cho, Huh, and Faber (2014) showed that strong sender trust influenced viral activity regardless of the level of brand trust. When sender trust is low, having a strong brand trust facilitates viral ad sharing activity. Shan and King (2015) also found that the brand relationship interacted with interpersonal ties in influencing consumers' decisions to pass along a eWOM message. Hayes and King (2014) report that stronger brand relationships positively influenced pass along, but did not consider influence of interpersonal relationships or UGC valence. Positive eWOM was shown to positively influence behavioural intentions while the reverse is true for negative eWOM (Doh and Hwang 2009; Jeong and Koo 2015). It is unclear how comment valence will intertwine with brand relationship and interpersonal relationship since a recipient must take into account the brand, the sender, and the comment when making a decision whether or not to accept and share a viral video ad in the SNS environment. The following hypothesis addresses this interconnected relationship.

H7: Brand relationship strength (BRS), interpersonal relationship strength (IRS), and comment valence interact to influence referral acceptance and re-share intention.

Methods

A 2 (BRS) × 2 (IRS) × 2 (user comment valence) online experiment was conducted to investigate the hypotheses. Each independent variable was manipulated and subsequently measured (1) to ensure manipulations were successful and (2) for analysis purposes; dependent variables (referral acceptance, re-share intention) and control variables (product involvement, product knowledge, source expertise (SE)) were measured. Real brands and interpersonal contacts were employed to increase external validity; further, brand familiarity is a necessary condition for brand satisfaction which, in turn impacts brand trust – the two components of a brand relationship (Ha and Perks 2005; Hess and Story 2005). Ad stimuli were created from viral videos ads that had previously gone viral.

The experiment was administered using Amazon Mechanical Turk (MTurk: www.Mturk. com). MTurk is a human intelligence platform whose participant compensation system streamlines online study design, recruitment of large participant pools, and data collection (Buhrmester, Kwang, and Gosling 2011). Berinsky, Huber, and Lenz's (2012) analysis suggests MTurk samples to outperform standard Internet samples and convenience samples in terms of demographic diversity and representativeness of the US population. As such, the platform has recently increased in used for academic research (e.g. Chu, Chen, and Sung 2016). Sessions lasted approximately 16 minutes.

Stimuli development

Pre-existing non-branded viral YouTube videos were edited into 30-second commercials. The edited videos were identical with the exception of brand identification occurring in the final four seconds of each stimulus. In accordance with *Advertising Age*'s criteria for viral status (see Cutler 2009), videos that had previously reached one million views were edited and pretested via online survey among 212 undergraduate students (71% female) in general elective courses in exchange for extra course credit. The ad stimulus corresponding to the computers category was selected as it exhibited adequate A_{Ad} ($M = 4.15$). Also, videos at least five years old were used to reduce the likelihood of participant recall of unedited videos.

Two additional pretests were conducted to identify appropriate stronger/weaker brand relationship pairs within the computer category and positive/negative user comment stimuli. Pretest 2 aimed to generate authentic user-generated comments that could be pretested for valence as well as to understand whether captions should comment on the video ad or the brand in the ad. Forty-seven undergraduate students at a large southeastern university were shown possible viral ad stimuli and asked to write captions they might add if sharing the video on Facebook. The exercise resulted in 10 user-generated comments to be further tested for appropriate valence as well as a clear indication that respondents comment on each the ad content and the brand in adding captions. Therefore, messages employed in the main study commented on each.

Pretest 3 surveyed 111 undergraduates in a general elective course at a large southeastern university in exchange for extra course credit. Following Hayes and King (2014), BRS was measured as a composite of brand satisfaction and brand trust (see also Hess and Story 2005; see Table 1 for major factor scales and alphas). Apple ($M = 5.52$) and Acer ($M = 3.67$) were identified as an appropriate stronger–weaker brand pairing ($t(110) = 10.07, p < .001$). Noteworthy is that familiar brands were employed; as brand relationships develop over a series of interactions, real brands were necessary to manipulate stronger and weaker relationships. Positively and negatively valenced user comments were also identified. On a seven-point bipolar scale, 'This video is great. Brand X is great' was rated the most positive comment ($M = 2.65$) while 'This video is terrible. Brand X is terrible' was rated most negative ($M = 5.43$). Note that mean scores indicate that floor and ceiling effects are avoided.

For each stimulus video, Facebook newsfeed notifications were produced within Facebook for authenticity (see Figure 1). Notifications were comprised of a valenced user comment, a thumbnail of the video, and a standardized video title.

Participants

In the main study, a national sample of US Facebook users aged 18–34 ($M = 28, N = 405$, 51% female) were recruited to participate in the online experiment. Participants were drawn from MTurk's participant pool and paid a $1.00 cash incentive based upon session length. Sixty-six per cent of participants were Anglo Americans ($n = 268$) followed by Asian Americans (10.4%; $n = 42$), Hispanic Americans (9.4%; $n = 38$), African Americans (8.4%; $n = 34$), and multiracial/others (5.6%; $n = 23$).

Procedure

Participants were directed to the online questionnaire after opting into the study. Following qualifying questions, seven-point semantic differential scales were used to measure two control variables: product involvement and product knowledge with regards to the computer category (see Table 1; Flynn and Goldsmith 1999; Zaichkowsky 1986). Prior research has shown that the influence of advertising messages on consumer attitudes

Table 1. Major factor scales and alphas.

Factors	Items	Source
Brand relationship strength		
Brand satisfaction (α = .85–.88)	I feel I know what to expect from Brand X. I am usually (feel that I would be) satisfied with Brand X products. I am usually (feel that I would be) satisfied with my experience with Brand X.	Hess and Story (2005)
Brand trust (α = .95–.96)	I trust Apple to offer me new products I may need I trust that Apple is interested in my satisfaction as a consumer Apple values me as a consumer of its product I trust Apple to offer me recommendations and advice on how to make the most of its product Apple offers me computers with a constant level of quality I trust that Apple will help me solve any problem I could have with the product.	Delgado-Ballester and Munuera-Alemán (2001)
Interpersonal relationship strength		
Interpersonal commitment (α = .86–.89)	I am committed to maintaining my relationship with Person X. I want our relationship to last for a long time. I feel very strongly linked to Person X. It is likely that I will delete this person as a Facebook friend within the next year.* I would not feel very upset if our relationship were to end in the near future.* I want our relationship to last forever. I am oriented toward continuing this relationship long-term. Our relationship is likely to end in the near future.*	Rusbult, Martz, and Agnew (1998)
Dyadic trust (α = .86–.93)	Person X is primarily interested in his (her) own welfare.* There are times when Person X cannot be trusted.* Person X is perfectly honest and truthful with me. I feel that I can trust Person X completely. Person X is truly sincere in his (her) promises. I feel that Person X does not show me enough consideration.* Person X treats me fairly and justify. I feel that Person X can be counted on to help me.	Larzelere and Huston (1980)
Ad referral acceptance (α = .92)	Unlikely/likely Improbable/probable Doubtful/doubtless Unpromising/promising	Sohn (2009) (adapted)
Intention to re-share (α = .91)		
Product involvement (α = .94)	Unimportant – Important Irrelevant – Relevant Mean nothing to me – Mean a lot to me Worthless – Valuable Boring – Interesting Unexciting – Exciting Unappealing – Appealing Mundane – Fascinating Not needed – Needed Not Involving – Involving	Zaichkowsky (1986)
Product knowledge (α = .92)	I know how to judge the quality of computers. I think I know enough about computers to feel pretty confident when I am purchasing I do not feel knowledgeable about computers* Among my circle of friends, I'm one of the 'experts' on computers Compared to most other people, I know less about computers* I have heard of most of the new computers that are around When it comes to computers, I really don't know a lot.* I can tell if computers are worth the price or not. I pretty much know about computers	Flynn and Goldsmith (1999)
Perceived source expertise (α = .96)	Not an expert – Expert Inexperienced – Experienced Unknowledgeable – Knowledgeable Unqualified – Qualified Unskilled – Skilled	Ohanian (1990)

*Reverse coded items.

150

a. Positive valence

This video is great! Apple is great!

 Apple However You Communicate

The video is great! Acer is great!

 Acer However You Communicate

b. Negative valence

This video is terrible! Apple is terrible!

 Apple However You Communicate

The video is terrible! Acer is terrible!

 Acer However You Communicate

Figure 1. Facebook news feed notifications.

and behaviours depends on consumers' level of product involvement (e.g. Dean et al. 1971; Gorn 1982; Yilmaz et al. 2011). Consumers put more cognitive effort into analysing messages when involvement is high, which increases attitudinal and behavioural change (Petty, Cacioppo, and Schumann 1983; Putrevu and Lord 1994). Another important influencing factor in eWOM research is consumers' level of knowledge about a product. Consumers with high versus low product knowledge tend to be more sceptical towards advertising claims, evaluate WOM credibility differently (do Paco and Reis 2012; Yilmaz et al. 2011), recall and retrieve more brand information for decision-making (Mitchell and Dacin 1996), and share more product-related information with others (Katz and Lazarsfeld 1955; Hennig-Thurau and Walsh 2003). Product involvement and knowledge effects are tested in the current research.

Next, two blocks of questions measured participants' existing attitude towards the brand (A_{Brand}) and BRS for Apple and Acer. BRS constituted a composite score of brand satisfaction and brand trust (Delgado-Ballester and Munuera-Alemán 2001; Hess and Story 2005). Brand question blocks were counterbalanced to reduce bias. To manipulate stronger and weaker interpersonal relationships, participants were asked to identify a close friend and a casual acquaintance within their online social network based upon appropriate levels of repeated interaction, reciprocity, and emotional attachment (Hinde 1995). Composite scores of interpersonal commitment and dyadic trust scales were measured and computed for each constituting IRS (Larzelere and Huston 1980; Rusbult, Martz, and Agnew 1998).

Next, subjects were randomly and evenly assigned to one of eight BRS–IRS–valence conditions. Within each condition, the corresponding Facebook referral notification was presented asking participants to indicate their likelihood of clicking on the video (referral acceptance) upon referral from the stronger/weaker IRS contact via Facebook newsfeed. The source of the message each participant saw was drawn from the names they provided

Table 2. Descriptive statistics and correlations between independent variables.

Variable	M	SD	1	2
1. BRS	4.88	1.32	–	
2. IRS	5.08	1.36	−.04	–
3. Valence	3.93	2.18	.02	−.04

Notes: BRS: brand relationship strength; IRS: interpersonal relationship strength.

in the questionnaire at the beginning of the study. Either the stronger or the weaker IRS, contact name was inserted in the post as the source of the video referral. Subsequently, subjects also rated their perception of the contact's expertise in the category (SE; Ohanian 1990). Since high expertise sources tend to be more persuasive in generating positive attitudes and behaviour change (e.g. Haung and Chen 2006), controlling for perceived SE allows interpersonal relationship effects to be isolated. Participants were then asked to watch the embedded ad stimulus; intention to re-share the ad shared by the contact was then assessed. Perceived valence of the user-generated caption was then measured on a seven-point scale from 'very negative' to 'very positive'. Table 2 reports means and standard deviations for correlations between variables.

Analysis and results

Paired-samples t-tests indicated appropriate statistically significant BRS differences: Apple ($M = 5.23$) – Acer ($M = 4.45$; $t(404) = 8.760$, $p < .001$). Identified stronger interpersonal relationships ($M = 6.01$) were also significantly stronger than weaker relationships ($M = 4.10$; $t(404) = 28.362$, $p < .001$). There was also appropriate statistically significant difference between positive ($M = 5.15$) and negative user comments ($M = 2.72$; $t(403) = 14.611$, $p < .001$). Manipulations were successful.

After mean centring all variables, multiple regression analyses were used to examine hypotheses and the research question. Tables 3 and 4 report regression coefficients for respective referral acceptance and re-share intention analyses. The control variables, product involvement ($M = 6.29$), and product knowledge ($M = 5.26$) showed no significant relationship with RA or RI; therefore, these variables were excluded from final models. Perceived SE ($M = 4.00$) exhibited strong significant relationships with acceptance ($R^2 = .177$) and re-share ($R^2 = .032$); thus, SE was controlled for in all models.

Hierarchical multiple regression was used to examine proposed main effects of independent variables on dependent variables. IRS had a strong, significant relationship with referral acceptance contributing an additional 9.0% of explained variance when added to the model with the control variable (SE); the overall model accounted for 26.7% of RA variance. However, IRS had no significant impact on re-sharing intention. BRS illustrated significant relationships with referral acceptance ($R^2 = .012$) and re-sharing ($R^2 = .018$). Lastly, no significant relationship was found between user comment valence (VAL) and RA; however, VAL did significantly influence re-share intention accounting for 2.6% change in r-square change when added to the model with SE. The model explained 5.8% of RI variance. Thus, H1, H3, H4, and H6 were supported; H2 and H5 were not supported.

Hierarchical multiple regression was employed to investigate the intertwining of BRS, IRS, and VAL in impacting referral acceptance and re-share intention (H7). When considered in the same model, BRS, IRS, and VAL exhibited significant relationships with referral acceptance. This is particularly interesting since valence was not a significant predictor

Table 3. Regression analyses for independent and control variables on ad referral acceptance.

Independent variables	B	SE B	β	t	Tolerance	VIF
IRS*	.445	.063	.328	7.025	.835	1.197
BRS*	.152	.069	.109	2.192		
Valence	.071	.038	.084	1.853	.996	1.004
Control variables						
Product involvement	.059	.102	.029	.575		
Product knowledge	−.003	.081	−.002	−.032		
Source expertise*	.484	.052	.421	9.317		
Interaction terms						
Constant				.123		
BRS				1.954	.972	1.029
IRS*	.010	.078		7.413	.824	1.214
Valence*	.116	.060	.083	1.975	.971	1.030
BRS × IRS*	.465	.063	.343	2.202	.982	1.018
Weaker IRS	.071	.036	.084	.657		
Stronger IRS*	.088	.040	.093	3.518		
BRS × VAL	.056	.085	.040	−.770	.954	1.049
IRS × VAL	.292	.083	.209	−.320	.988	1.012
BRS × IRS × VAL*	−.020	.026	−.033	−2.314	.969	1.032
Pos–Weaker IRS*	−.008	.025	-.014	2.225		
Pos– Stronger	−.040	.017	−.099	3.696		
IRS*	.266	.120	.196	−1.022		
Neg– Weaker IRS	.402	.109	.296	1.272		
Neg– Stronger	−.120	.117	−.084			
IRS	.155	.122	.109			

Notes: All data based upon centred variables; *significant such that $p < .05$.

when analysed alone (H5). Further, two significant interactions emerged: BRS × IRS and BRS × IRS × VAL. Following Aiken, West, and Reno (1991), interactions were probed by examining regression models at −1 standard deviation (low) and +1 standard deviation (high) from the moderator mean (see also Irwin and McClelland 2003). Findings indicate that BRS had no significant influence on RA when the referral is made by a weaker IRS referrer but a strong, significant influence when a stronger IRS referrer shares the ad (see Figure 2). A comparison of slopes from each model supports this interpretation ($z = 2.440$, $p = .007$).

Table 4. Regression analyses for independent and control variables on re-share intention

Independent variables	B	SE B	β	t	Tolerance	VIF
IRS	−.093	.073	−.068	−1.274	.835	1.197
BRS*	.190	.069	.136	2.751		
Valence*	.137	.041	.162	3.336	.996	1.004
Control Variables						
Product Involvement	−.045	.102	−.022	−.440		
Product Knowledge	.066	.081	.040	.808		
Source Expertise*	.206	.056	.180	3.664		
Interactions						
Constant	.010	.089		.112		
BRS*	.139	.068	.099	2.052	.972	1.029
IRS	−.065	.071	−.048	−.905	.824	1.214
Valence*	.116	.041	.137	2.822	.971	1.030
BRS X IRS	.072	.046	.077	1.588	.982	1.018
BRS X VAL*	.059	.030	.097	1.976	.954	1.049
Positive*	.307	.093	.219	3.318		
Negative	.054	.097	.039	.561		
IRS X VAL	.054	.029	.091	1.895	.988	1.012
BRS X IRS X VAL	−.023	.020	−.057	−1.167	.969	1.032

Notes: All data based upon centered variables; *significant such that $p < .05$.

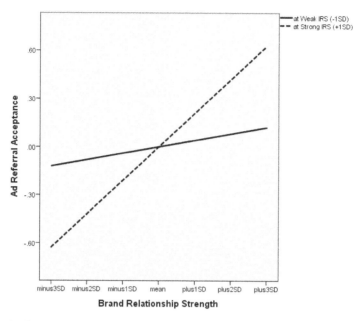

Figure 2. Simple slopes analysis of the interaction between brand relationship strength and interpersonal relationship strength on ad referral acceptance.

Analysis of the three-way BRS × IRS × user comment valence interaction provides insight into the role of comment valence. Two separate simple slope analyses were performed: BRS × IRS when VAL was positive, BRS × IRS when VAL was negative. Examining the BRS × IRS interaction when VAL was positive illustrated that BRS had a significant influence regardless of IRS (see Figure 3; $z = 1.1028$, $p = 0.1351$). However, when VAL was negative, BRS influence was negated regardless of stronger versus weaker IRS.

Finally, analysis of combined effect of the independent variables on re-share intention revealed a BRS X VAL interaction. Simple slopes analysis indicated that BRS had no significant influence on intention to re-share the ad when the comment was negative but a strong, significant impact when the comment was positive ($z = 2.5995$, $p = .005$). Figure 4 depicts the interaction.

General discussion

Summary and implications

Viral advertising is a new hybrid of traditional advertising and eWOM that is not yet fully understood by researchers and practitioners. Its unique ability to leverage the more trusted and influential endorsement of personal networks while maintaining message origination with the brand makes viral advertising a potentially powerful tool for brands. Understanding the influence of the brand and interpersonal contacts that serve as referrers is crucial. This study extends a developing area of research in viral advertising that examines how interpersonal and brand relationships intertwine to impact viral sharing decisions by inserting user comment valence into the equation. The results also show that

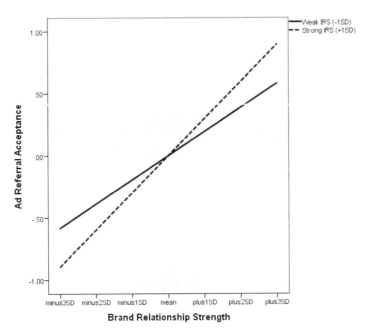

Figure 3. Simple slopes analysis of the interaction between brand relationship strength and interpersonal relationship strength on ad referral acceptance when UGC valence is positive.

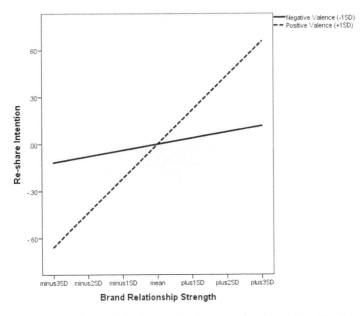

Figure 4. Simple slopes analysis of the interaction between brand relationship strength and UGC valence on ad re-share intention.

interpersonal relationships, brand relationships, and user comment valence play different but intertwining roles in SNS users' acceptance of viral video ads and decisions to re-share.

For referral acceptance, brand and interpersonal relationships help to influence recipients of viral ads to accept referrals. However, the interpersonal relationship has far stronger influence. In the final model, brand relationship main effects were overpowered by interpersonal relationship impact. Examination of the interaction between BRS and IRS revealed that the brand's impact increases as IRS increases suggesting that recipients consider the interpersonal contact first and then the brand. As discussed earlier, relationships are constructed and maintained through a series of net positive interactions leading to trust and, then, commitment to the relationship for the purpose of reducing uncertainty in exchange and due to positive emotion developed between partners (Cook and Emerson 1984; Kollock 1994; Lawler, Thye, and Yoon 2000). Decisions to exchange are cognitive as well as affective. In line with this, Fang (2014) drew from the stimulus–organism–response model and affect-as-information theory to illustrate that affective (e.g. arousal) as well as cognitive (e.g. perceived credibility) information about the source factors into eWOM adoption decisions. According to affect-as-information theory, more relevant affect-based information will carry more weight in judgement (e.g. Pham 1998). In the context of social networking sites, social interaction with other people is a primary motivation (Chen 2013); therefore, it is not surprising that recipients look first to the interpersonal contact and then to the brand when making decisions.

Further, the bolstering of brand influence when ads are shared by stronger interpersonal contact is in line with extant marketing literature regarding spillover effects (e.g. Ahluwalia, Unnava, and Burnkrant 2001). Based on information integration theory, the premise of spillover effects is that attitudes and beliefs about brands are constantly modified as more stimulus information is integrated with existing attitudes and beliefs. More salient cues will be more likely to be accessed and bias information processing (Houston and Fazio 1989; Fazio 1986; Fazio 1989; Fazio and Williams 1986). Further, brand judgements are influenced by characteristics of material in close proximity to the brand (Lynch, Chakravarti, and Mitra 1991). Simonin and Ruth (1998) illustrate this principle in the context of brand alliances, wherein two brands partner in a marketing communications context. The study showed that (1) brand alliances resulted in significant spillover effects for each partner brand, and (2) the more salient brand exerted greater effect on the less salient partner. Sharing ads can be characterized as an alliance between the ad's object brand and the interpersonal contact referring the ad. In the context of SNSs, the interpersonal contact is the more salient partner as social interaction with contacts is a primary driver of usage. As such, it follows that brand influence would be strengthened as a result of spillover effects via association with stronger interpersonal contacts. This finding not only clearly illustrates the much lauded advantage of seeding eWOM through personal contacts (e.g. Mangold and Faulds 2009), but also implies that brands can acquire new positive associations and mitigate existing negative associations when brand content is shared. This idea is ripe for future research that explores not only how associations may be traded between brands and interpersonal contacts but also the permanence of such associations and their implications for consumer–brand and interpersonal relationships.

The widely accepted practice of targeting eMavens for initial seeding of hopeful viral campaign may be called into question as a result of the present findings. While eMavens are highly extraverted gathering and disseminating high volumes of information (e.g. Wasko and Faraj 2005), the present findings suggest that targeting tighter knit groups may be more beneficial for referral acceptance. Further, as the brand becomes more influential when close interpersonal ties are involved, facilitating positive interactions with contacts within more dense networks such as Facebook on a regular basis offers brands relationship building opportunities whereas passing messages through weak ties (e.g. eMavens) is likely better for awareness (Hayes and King 2014).

Adding valence into the equation uncovered an interesting caveat, when a positive user comment was attached to the ad referral, BRS had significant impact regardless of the interpersonal contact; conversely, when a negative user comment was attached, the brand was not impactful irrespective of IRS. This finding is in line with previous research on negativity effect suggesting that negative information more strongly influences impression formation than positive information because it is scarcer and more diagnostic (Chiou and Cheng 2003; Herr et al. 1991; Lee et al. 2008; Xue and Zhou 2010; Yang and Mai 2010). Indeed, in an analysis of consumer posts on SNS brand pages, De Vries, Gensler, and Leeflang (2012) found only 11% of posts to be negative; further, negative eWOM has shown to have stronger influence than positive eWOM (Jeong and Koo 2015).

This finding highlights the importance of actively building strong brand relationships with consumers online. Trusted brands garner positive brand affect as well as purchase and attitudinal loyalty (Chaudhuri and Holbrook 2001). Further, Ranaweera and Prabhu (2003) illustrate brand satisfaction and brand trust to be strong predictors of positive word-of-mouth. Positive eWOM positively impacts behavioural intention (Doh and Hwang 2009). While negative information often outweighs positive information, the cue–diagnosticity model holds that a positivity effect can occur when positive eWOM is more diagnostic than negative eWOM (Skowronski and Carlston 1989). Therefore, brands should work to create continued positive interactions with their consumers and encourage them to make positive posts via various forms of eWOM; meanwhile, negative eWOM effects should be limited through timely response to negative eWOM (see also Jeong and Koo 2015). More positive and useful user comments will facilitate greater referral acceptance. Also shown as important is the continued production of brand content considered valuable and shareworthy for the immediate brand consumer as well as their social networks as negative evaluations can quickly halt the viral spread (see also Hayes and King 2014).

Re-share intention is influenced by brand relationship and user comment valence. Findings regarding brand relationship are consistent extant research. Hayes and King (2014) report that brand relationship factors in the referral process as strong consumer–brand relationship supplement the interpersonal relationship in influencing decisions to share a viral video ad. Consistent with SET, the findings indicate that positive user-generated comments increase the intention to re-share a viral video ad wherein viral advertising referrers seek mutually beneficial exchanges when positive outcomes are expected. Previous research has provided empirical evidence that the likelihood of sharing an ad is influenced by reciprocal altruism and expected positive outcomes (Hayes and King 2014; Hayes, King, and Ramirez 2016). This study extended the current literature on eWOM by identifying the impact of user comment valence on intention to re-share. The results show that brand relationship has a strong positive impact on re-share intention when user-generated comment is positive. However, when the comment is negative, brand relationship has no impact on intention to re-share. This finding again highlights the importance of brands cultivating relationships with target markets online. Stronger brand relationships guard against negative word-of-mouth engendering positive word-of-mouth (Chaudhuri and Holbrook 2001) facilitating acceptance and re-sharing of viral ads.

The differences in how the independent variables factored into referral acceptance decisions versus re-share decisions are noteworthy. The interpersonal contact weighed heavily into referral acceptance but not re-share intention. This is in line with previous research. Hayes, King, and Ramirez (2016) reported interpersonal and brand relationships intertwining to influence referral acceptance of viral ads. Hayes and King (2014) proposed and tested a model of co-referral of viral ads (equivalent to re-share intention) wherein only the brand relationship along with sharing motivations factored into co-referral; results suggested that re-sharing of ads is a relational maintenance function wherein the potential re-sharer must access the perceived value of the brand content for their social network. The referral acceptance decision, conversely, is an assessment only of the content's value for that individual wherein the only information available for assessment is the brand, the interpersonal contact referring the ad, and perhaps an attached user comment; therefore, the interpersonal contact influences referral acceptant but not re-share intention.

The intertwining roles of interpersonal relationships, brand relationships, and comment valence in viral sharing decisions further illustrate relational construction and maintenance function of brand content sharing as social exchange. SNS users perform cost–benefit analyses when deciding to accept and re-share ads. Further, relationships are developed through positive interactions. Present results show that indeed positive interactions lead to subsequent positive interactions. However, the halting power of negative interactions represents an important threat to brand content sharing and brand relationships. Molm (1990) notes that, just as positive social exchange leads to satisfaction, trust, and commitment to relationships, negative exchanges deteriorate and even end relationships. The negativity effect illustrated in these findings suggests that negative interactions may be more destructive than positive interactions are destructive. Viral advertising research must understand, then, exactly how and to what extent both positive and negative word-of-mouth within the sharing process impact brand relationships, while marketers must actively manage positive and negative eWOM.

Limitations and future research

As with all studies, the findings of this research have limitations. First, participants for the study were adult Facebook users aged 18–34. While appropriate for this study, future research should replicate the findings across age groups and platforms.

A difficulty of research in the SNS domain is gaining access necessary to observe actual sharing behaviour rather than self-report data as used in this study. Ideally, collaboration with industry to allow a longitudinal field experiment (e.g. Cho, Huh, and Faber 2014) would allow for a more realistic examination of behaviour across product categories and varying level of positive and negative message valences.

Videos that had previously gone viral were edited and used as the basis for new ad stimuli. While it was necessary to identify video stimuli worth sharing, novelty is an important factor in the decision to share an ad (Hennig and Phillips 2012). If the edited stimuli seemed familiar to the participants, ad acceptance and re-share intention may have been decreased. However, it should be noted that such a case would not influence the effects of independent variables.

Due to the importance of using real brands in the study, it is possible the differences in brand awareness, a variable not measured, could bias results. While not measured directly, pre-tests did check for adequate brand attitude levels for each Apple and Acer. Brand awareness is requisite for brand attitude formation (Keller 1993) which, in turn, is encompassed within the brand relationship independent variable (Hess and Story 2005). Therefore, this concern is mitigated from a theoretical standpoint; still, future studies should directly measure and account for possible brand awareness effects.

Despite limitations, the insights provided are valuable to the understanding of how brands and friends intertwine in the social processes involved in viral advertising sharing. Brand and interpersonal relationships are both integral parts of 'going viral.' Distinct viral processes are investigated. This study provides a first look into these issues.

Acknowledgment

This research was funded by the Jim Kennedy New Media Professorship Fund. A portion of the research was conducted at the Zimmerman School of Advertising & Mass Communications at the University of South Florida.

Disclosure statement

No potential conflict of interest was reported by the authors.

References

Ahluwalia, R., H. R. Unnava, and R. E. Burnkrant. 2001. The moderating role of commitment on the spillover effect of marketing communications. *Journal of Marketing Research* 38, no. 4: 458–470.

Aiken, L., S. G. West, and R. R. Reno. 1991. *Multiple regression: Testing and interpreting interactions.* Thousand Oaks, CA: Sage.

Anderson, E. W., and L. C. Salisbury. 2003. The formation of market-level expectations and its covari-ates. *Journal of Consumer Research* 30, no. 1: 115–124.

Bhattacharyya, S. 2016. Shell's clean energy campaign tops viral videos charts for 2016. http://adage.com/article/the-viral-video-chart/top-10-campaigns-2016/307190/

Berger, J., A. T. Sorensen, and S. J. Rasmusseen. 2010. Positive effects of negative publicity: When negative reviews increase sales. *Marketing Science* 29, no. 5: 815–827.

Berinsky, A. J., G. A. Huber, and G. S. Lenz. 2012. Evaluating online labour markets for experimental research: Amazon.com's mechanical Turk. *Political Analysis* 20, no. 3: 351–368.

Blau, P. M. 1964. *Exchange & Power in Social Life*. New York: John Wiley & Sons, Inc.

Bourdieu, P. 1984. *Distinction: A Social Critique of the Judgment of Taste*. Cambridge, MA: Harvard University Press.

Brown, D. E. 1991. *Human Universals*. Philadelphia, PA: Temple University Press.

Buhrmester, M., T. Kwang, and S. D. Gosling. 2011. Amazon's mechanical Turk: A new source of inex-pensive, yet high-quality, data? *Perspectives on Psychological Science* 6, no. 1: 3–5.

Campbell, C., L. F. Pitt, M. Parent, and P. R. Berthon. 2011. Understanding consumer conversations around ads in a Web 2.0 World. *Journal of Advertising* 40, no. 1: 87–102.

Chaudhuri, A., and M. B. Holbrook. 2001. The chain of effects from brand trust and brand affect to brand performance: The role of brand loyalty. *Journal of Marketing* 65, no. 2: 81–93.

Chen, R. 2013. Member use of social networking sites – an empirical examination. *Decision Support Systems* 54, no. 3: 1219–1227.

Cheung, M. Y., C. Luo, C. L. Sia, and H. Chen. 2009. Credibility of electronic word-of-mouth: Informa-tion and normative determinants of on-line consumer recommendations. *International Journal of Electronic Commerce* 13, no. 4: 9–38.

Chevalier, J. A., and D. Mayzlin. 2006. The effect of word of mouth on sales: Online book reviews. *Journal of Marketing Research* 43, no. 3: 345–354.

Chiou, J., and C. Cheng. 2003. Should a company have message boards on its web site? *Journal of Interactive Marketing* 17, no. 3: 50–61.

Chiu, H.-C., Y.-C. Hsieh, Y.-H. Kao, and M. Lee. 2007. The determinants of email receivers' disseminat-ing behaviours on the internet. *Journal of Advertising Research* 47, no. 4: 524–534.

Cho, S., J. Huh, and R. J. Faber. 2014. The influence of sender trust and advertiser trust on multistage effects of viral advertising. *Journal of Advertising* 43, no. 1: 100–114.

Chu, S.-C., and Y. Kim. 2011. Determinants of consumer engagement in electronic word-of-mouth (eWOM) in social networking sites. *International Journal of Advertising* 30, no. 1: 47–75.

Chu, S.-C. 2011. Viral advertising in social media: Participation in Facebook groups and responses among college-aged user. *Journal of Interactive Advertising* 12, no. 1: 30–43.

Chu, S.-C., H.-T. Chen, and Y. Sung. 2016. Following brands on Twitter: An extension of theory of planned behaviour. *International Journal of Advertising* 35, no. 3: 421–437.

Constant, D., S. Kiesler, and L. Sproull. 1994. What's mine is ours, or is it? A study of attitudes about information sharing. *Information Systems Research* 5: 400–421.

Cook, K. S., and R. M. Emerson. 1984. Exchange networks and the analysis of complex organizations. *Research in the Sociology of Organizations* 3, no. 4: 1–30.

Cutler, M. 2009. How to make your online video go viral. *Advertising Age* 42, March 30.

De Vries, L., S. Gensler, and P. S. H. Leeflang. 2012. Popularity of brand posts on brand fan pages: An inves-tigation of the effects of social media marketing. *Journal of Interactive Marketing* 26, no. 2: 83–91.

Dean, R. B., J. A. Austin, and W. A. Watts. 1971. Forewarning effects in persuasion: Field and classroom experiments. *Journal of Personality and Social Psychology* 18, no. 2: 210–221.

Delgado-Ballester, E., and J. L. Munuera-Alemán. 2001. Brand trust in the context of consumer loyalty. *European Journal of Marketing* 35, no. 11/12: 1238–1258.

do Paco, A. M. F., and R. Reis. 2012. Factors affecting scepticism toward green advertising. *Journal of Advertising* 41, no. 4: 147–155.

Doh, S. J., and J. S. Hwang. 2009. How consumers evaluate eWOM (electronic word-of-mouth) mes-sages. *CyberPsychology & Behavior* 12, no. 2: 193–197.

Duan, W., B. Gu, and A. B. Whinston. 2008. Do online reviews matter? An empirical investigation of panel data. *Decision Support System* 45, no. 5: 1007–1016.

Eckler, P., and P. Bolls. 2011. Spreading the virus: Emotional tone of viral advertising and its effect on forwarding intentions and attitudes. *Journal of Interactive Advertising* 11, no. 2: 1–11.

eMarketer. 2015. Social network ad spending to hit $23.68 billion worldwide in 2015. http://www.emarketer.com/Article/Social-Network-Ad-Spending-Hit-2368-Billion-Worldwide-2015/1012357

eMarketer. 2016a. Marketers expected to shift more dollars toward desktop, mobile video ads. http://www.emarketer.com/Article/Marketers-Expected-Shift-More-Dollars-Toward-Desktop-Mobile-Video-Ads/1013959.

eMarketer. 2016b. US marketers to step up social video ads, particularly on Facebook. https://www.emarketer.com/Article/US-Marketers-Step-Up-Social-Video-Ads-Particularly-on-Facebook/1014283.

Fang, Y.-H. 2014. Beyond the credibility of electronic word of mouth: Exploring eWOM adoption on social networking sites from affective and curiosity perspectives. *International Journal of Electronic Commerce* 18, no. 3: 67–102.

Fazio, R. H. 1986. How do attitudes guide behavior? In *The Handbook of Motivation and Cognition: Foundations for Social Behavior*, ed. R. M. Sorrentino and E. T. Higgins, 204–243. New York: Guilford Press.

Fazio, R. H. 1989. On the power and functionality of attitudes: The role of attitude accessibility. In *Attitude Structure and Function*, ed. A. Pratkanis, S. Breckler, and A. Greenwald, 153–179. Hillsdale, NJ: Lawrence Erlbaum Associates.

Fazio, R. H., and C. J. Williams. 1986. Attitude accessibility as a moderator of the attitude-perception and attitude-behavior relations: An investigation of the 1984 presidential elections. *Journal of Personality and Social Psychology* 51, September: 505–514.

Fishbein, M., and I. Ajzen. 1975. *Beliefs, attitudes, intentions and behaviour: An introduction to theory and research*. Reading, MA: Addition-Wesley.

Flynn L. R., and R. E. Goldsmith. 1999. A short reliable measure of subjective knowledge. *Journal of Business Research* 46: 57–66.

Fournier, S. 1998. Consumers and their brands: Developing relationship theory in consumer research. *Journal of Consumer Research* 21, no. 4: 343–353.

Fournier, S., and C. Alvarez. 2012. Brands as relationship partner: Warmth, competence, and in-between. *Journal of Consumer Psychology* 22: 177–185.

Frenzen, J., and K. Nakamoto. 1993. Structure, cooperation, and the flow of market information. *Journal of Consumer Research*: 360–375.

Godes, D., and D. Mayzlin. 2004. Using online conversations to study word-of-mouth communications. *Marketing Science* 23, no. 4: 545–560.

Gorn, G. J. 1982. The effects of music in advertising on choice behaviour: A classical conditioning approach. *Journal of Marketing* 46, no. 1: 94–101.

Granovetter, M. S. 1973. The strength of weak ties. *American Journal of Sociology* 78, no. 6: 1360–1380.

Ha, H.-Y., and H. Perks. 2005. Effects of consumer perceptions of brand experience on the web: Brand familiarity, satisfaction and brand trust. *Journal of Consumer Behaviour* 4, no. 6: 438–452.

Harrison-Walker, J. L. 2001. The measurement of word-of-mouth communication and an investigation of service quality and customer commitment as potential antecedents. *Journal of Service Research* 4, no. 1: 60–75.

Hayes, J. L. 2016. Building relationships with empowered consumers. In *The New Advertising: Branding, Content, and Consumer Relationships in the Data-driven Social Media Era*, ed. R. E. Brown, V. K. Jones, and M. Wang. Santa Barbara, CA: Praeger ABC-Clio.

Hayes, J. L., and K. W. King. 2014. The social exchange of viral ads: Referral and coreferral of ads among college students. *Journal of Interactive Advertising* 14, no. 2: 98–109.

Hayes, J. L., K. W. King, and A. Ramirez. 2016. Brands, friends, & viral advertising: A social exchange perspective on the ad referral processes. *Journal of Interactive Marketing* 36: 31–45.

Hennig, D., and B. J. Phillips. 2012. Understanding viral video mavens. Presented at the 2012 American Academy of Advertising Annual Conference, March, Myrtle Beach, SC.

Hennig-Thurau, T., and G. Walsh. 2003. Electronic word-of-mouth: Motives for and consequences of reading customer articulations on the internet. *Interpersonal Journal of Electronic Commerce* 8, no. 2: 51–72.

Herr, P. M., F. R. Kardes, and J. Kim. 1991. Effects of word-of-mouth and product-attitude information on persuasion: An accessibility-diagnosticity perspective. *Journal of Consumer Research* 17, no. 4: 454–62.

Hess, J., and J. Story. 2005. Trust-based commitment: Multidimensional consumer-brand relationships. *Journal of Consumer Marketing* 22, no. 6: 313–322.

Hinde, R. A. 1995. A suggested structure for a science of relationships. *Personal Relationships* 2, no. 1: 1–15.

Houston, D. A., and R. H. Fazio. 1989. Biased processing as a function of attitude accessibility: Making objective judgments subjectively. *Social Cognition* 7, Spring: 51–66.

Hsu, C.-L., and J. C. C. Lin. 2008. Acceptance of blog usages: The roles of technology acceptance, social influence and knowledge sharing motivation. *Information and Management* 45, no. 1: 65–74.

Huang, J. -H., and Chen Y.-F. 2006. Herding on online product choice. *Psychology and Marketing* 23: 413–428.

Irwin, J. R., and G. H. McClelland. 2003. Negative consequences of dichotomizing continuous predictor variables. *Journal of Marketing Research* 40, no. 3: 366–371.

Jarvenpaa, S., K. Knoll, and D. E. Leidner. 1998. Is anybody out there? Antecedents of trust in global virtual teams. *Journal of Management Information Systems* 14, no. 4: 29–64.

Jeong, H.-J., and D.-M. Koo. 2015. Combined effects of valence and attributes of e-WOM on consumer judgment for message and product the moderating effect of brand community type. *Internet Research* 25, no. 1: 2–29.

Katz, E., and P. F. Lazarsfeld. 1955. *Personal Influence: The Part Played by People in the Flow of Mass Communications*. New York: The Free Press.

Kaplan, K. J. 1972. One the ambivalence-indifferent problem in attitude theory and measurement: A suggested modification of the semantic differential technique. *Psychological Bulletin* 77, no. 5: 361–372.

Kaplan, A. M., and M. Haenlein. 2010. Users of the World, unite! The challenges and opportunities of social media. *Business Horizons* 53, no. 1: 59–68.

Keller, K. L. 1993. Conceptualizing, measuring, and managing customer-based brand equity. *Journal of Marketing* 57, no. 1: 1–22.

Kiyonari, T., S. Tanisa, and T. Yamagishi. 2000. Social exchanges and reciprocity: Confusion or a heuristic. *Evolution and Human Behavior* 21: 411–427.

Kollock, P. 1994. The emergence of exchange structures: An experimental study of uncertainty, commitment, and trust. *American Journal of Sociology* 100, no. 2: 313–345.

Larzelere, R. E., and T. L. Huston 1980. The dyadic trust scale: Toward understanding interpersonal trust in close relationships. *Journal of Marriage and the Family* 42, no. 3: 595–604.

Lawler, E. J., S. R. Thye, and J. Yoon. 2000. Emotion and group cohesion in productive exchange. *American Journal of Sociology* 106, no. 3: 616–657.

Lee, J., C.-D. Ham, and M. Kim. 2013. Why people pass along online video advertising: From the perspective of the interpersonal communication motives scale and the theory of reasoned action. *Journal of Interactive Advertising* 13, no. 1: 1–13.

Lee, M., and S. Youn. 2009. Electronic word of mouth (eWOM) how eWOM platforms influence consumer product judgement. *International Journal of Advertising* 28, no. 3: 473–499.

Lee, J., D. Park, and I. Han. 2008. The effect of negative online consumer reviews on product attitude: An information processing view. *Electronic Commerce Research and Applications* 7, no. 3: 341–352.

Lee, M., S. Rodgers, and M. Kim. 2009. Effects of valence and extremity of eWOM on attitude toward the brand and website. *Journal of Current Issues & Research in Advertising* 31, no. 2: 1–11.

Liu, Y. 2006. Word of mouth for movies: Its dynamics and impact on box office revenue. *Journal of Marketing* 70, no. 3: 74–89.

Lynch, J. G., D. Chakravarti, and A. Mitra. 1991. Contrast effects in consumer judgments: Changes in mental representations or in the anchoring of rating scales? *Journal of Consumer Research* 18, no. 3: 284–297.

Mangold, G. W., and D. J. Faulds. 2009. Social media: The new hybrid element of the promotion mix. *Business Horizons* 52, no. 4: 357–365.

Mitchell, A. A., and P. A. Dacin. 1996 The assessment of alternative measures of consumer expertise. *Journal of Consumer Research* 23, no. 3: 219–239.

Molm, L. D. 1990. Structure, action, and outcomes: The dynamics of power in social exchange. *American Sociological Review* 55, no. 3: 427–447.

Molm, L. D., N. Takahashi, and G. Peterson. 2000. Risk and trust in social exchange: An experimental test of a classical proposition. *American Journal of Sociology* 105, no. 5: 1396–1427.

Morgan, R. M., and S. D. Hunt. 1994. The commitment-trust theory of relationship marketing. *Journal of Marketing* 58, no. 3: 20–38.

Nan, X., and R. J. Faber. 2004. Advertising theory: Reconceptualising the building's blocks. *Marketing Theory* 4, no. 1–2: 7–30.

Ohanian, R. 1990. Construction and validation of a scale to measure celebrity endorsers' perceived expertise, trustworthiness, and attractiveness. *Journal of Advertising* 19, no. 3: 39–52.

Petty, R. E., J. T. Cacioppo, and David Schumann. 1983. Central and peripheral routes to advertising effectiveness: The moderating role of involvement. *Journal of Consumer Research* 10, no. 2: 135–146.

Pham, M. T. 1998. Representativeness, relevance, and the use of feelings in decision making. *Journal of Consumer Research* 25, no. 2: 144–159.

Phelps, J. E., R. Lewis, L. Mobilo, D. Petty, and R. Niranjan. 2004. Viral marketing or electronic word-of-mouth advertising: Examining consumer responses and motivations to pass along email. *Journal of Advertising Research* 44, no. 4: 333–348.

Porter, L., and G. J. Golan. 2006. From subservient chickens to brawny men. *Journal of Interactive Advertising* 6, no. 2: 4–33.

Putrevu, S., and K. R. Lord. 1994. Comparative and noncomparative advertising: Attitudinal effects under cognitive and affective involvement conditions. *Journal of Advertising* 23, no. 2: 77–91.

Ranaweera, C., and J. Prabhu. 2003. On the relative importance of customer satisfaction and trust as determinants of customer retention and positive word of mouth. *Journal of Targeting, Measurement and Analysis for marketing* 12, no. 1: 82–90.

Richins, M. L. 1983. Negative word-of-mouth by dissatisfied consumers: A pilot study. *Journal of Marketing* 47, no. 1: 68–78.

Rim, H., and D. Song. 2016. How negative becomes less negative: Understanding the effects of comment valence and response sidedness in social media. *Journal of Communication* 66: 475–495.

Rosenkrans, G. 2009. The creativeness and effectiveness of online interactive rich media advertising. *Journal of Interactive Advertising* 9, no. 2: 18–31.

Rusbult, C. E., J. M. Martz, and C. B. Agnew. 1998. The investment model scale: Measuring commitment level, satisfaction level, quality of alternatives, and investment size. *Personal Relationships* 5, no. 4: 357–387.

Shan, Y., and K. W. King. 2015. The effects of interpersonal tie strength and subjective norms on consumers' brand-related eWOM referral intentions. *Journal of Interactive Advertising* 15, no. 1: 16–27.

Simonin, B. L., and J. A. Ruth. 1998. Is a company known by the company it keeps? Assessing the spillover effects of brand alliances on consumer brand attitudes. *Journal of Marketing Research* 35, no. 1: 30–42.

Skowronski, J. J., and D. E. Carlston. 1989. Negativity and extremity biases in impression formation: A review of explanations. *Psychological Bulletin* 105, no. 1: 131–142.

Smith, T., J. R. Coyle, E. Lightfood, and A. Scott. 2007. Reconsidering models of influence: The relationship between consumer social networks and word-of-mouth effectiveness. *Journal of Advertising Research* 47: 387–397.

Sohn, D. 2009. Disentangling the effects of social network density on electronic word of mouth (eWom) intention. *Journal of Computer Mediated Communication* 14, no. 2: 352–367.

Tang, T., E. Fang, and F. Wang. 2014. Is neutral really neutral? The effects of neutral user-generated content on product sales. *Journal of Marketing* 78: 41–58.

Taylor, D., D. Strutton, and K. Thompson. 2012. Self-enhancement as a motivation for sharing online advertising. *Journal of Interactive Advertising* 12, no. 2: 13–28.

van Noort, G., Antheunis, M. L., and E. A. van Reijmersdal. 2012. Social connections and the persuasiveness of viral campaigns in social network sites: Persuasive intent as the underlying mechanism. *Journal of Marketing Communications* 18, no. 1: 39–53.

Wasko, M. M.., and S. Faraj. 2005. Why Should i share? Examining social capital and knowledge contribution in electronic networks of practice. *MIS Quarterly* 29, no. 1: 35–57.

Willemsen, L. M., P. C. Neijens, G. Bronner, and J. A. de Ridder. 2011. Highly recommended!" The content characteristics and perceived usefulness of online consumer reviews. *Journal of Computer-Mediated Communication* 17, no. 1: 19–38.

Xue, F., and P. Zhou. 2010. The effects of product involvement and prior experience on Chinese consumers' responses to online word of mouth. *Journal of International Consumer Marketing* 23, no. 1: 45–58.

Yang, J. and E. Mai. 2010. Experiential goods with network externalities effects: An empirical study of online rating system. *Journal of Business Research* 63, no. 9/10: 1050–1057.

Yilmaz, C. E., E. Telci, M. Bodur, and T. E. Iscioglu. 2011. Source characteristics and advertising effectiveness. *International Journal of Advertising* 30, no. 5: 889–914.

Zaichkowsky, J. L. 1986. Conceptualizing involvement. *Journal of Advertising* 15, no. 2: 4–34.

Zeng, M., and W. Reinartz. 2003. Beyond online search: The road to profitability. *California Management Review* 45, no. 2: 107–130.

Zhu, F., and X. Zhang. 2010. Impact of online consumer reviews on sales: The moderating role of product and consumer characteristics. *Journal of Marketing* 74, no. 2: 133–148.

Appendix 1: Description of ad stimuli

Ad stimuli videos featured a young man was his computer to make video of himself lip-syncing the song "Numa Numa." Particularly, the ad stimuli focus on the portion of the song wherein the young man sings "Hello" in multiple languages while waving to video viewers. The video then concludes with a brand logo shot (see Figure 1) and the tagline, "For However You Communicate."

Index

Note: Page numbers in bold and italics refer to tables and figures.

Milton Keynes UK
Ingram Content Group UK Ltd.
UKHW051854071024
449327UK00025B/1955